CONVERGING WORLD

John Pontin OBE set up JT Design Build in 1961. Over a period of 35 years the company has developed a national profile, in particular for its stance on green issues in construction. John has a keen interest in sustainability. Besides being the vice-president of the Schumacher Society, he sits on the boards of Sustainability West, Business West and the Royal Society of Arts Advisory Council. He was Chair of the Dartington Hall Trust from 1984 to 1997, and a board member of the Natural Step (part of Forum for the Future). He was awarded an Honorary Doctor of Laws by the University of Bristol in 2007.

Ian Roderick is the Director of the Schumacher Institute for Sustainable Systems, a 'think and do tank' created by the Schumacher Society. He is also president of the UK Systems Society and a leading light in the Go Zero project in the village of Chew Magna. His early career was in Operational Research, and he co-founded a software company in 1982. In 2002 he completed an MSc in Responsibility & Business Practice at Bath University; he is also a Fellow of the Royal Society of Arts.

John and Ian both live in Chew Magna, a village just south of the British city of Bristol.

Schumacher Briefing No. 13

CONVERGING WORLD

Connecting Communities in Global Change

John Pontin & Ian Roderick

Foreword by Cletus Babu

published by Green Books
for The Schumacher Society

First published in 2007
by Green Books Ltd
Foxhole, Dartington, Totnes,
Devon TQ9 6EB
www.greenbooks.co.uk

for The Schumacher Society
The CREATE Centre, Smeaton Road,
Bristol BS1 6XN
www.schumacher.org.uk
admin@schumacher.org.uk

Printed by MPG Books, Bodmin, Cornwall, UK

Text printed on Corona Natural (100% recycled)
Covers and colour plates printed on GreenCoat Velvet (80% recycled)

A catalogue record for this publication
is available from the British Library

ISBN 978 1 903998 92 2

The Schumacher Briefings
Series Editor: Stephen Powell
Founding Editor: Herbert Girardet

Contents

The Converging World

'Connecting communities through action
to maximise our quality of life
while achieving equal and minimal impact
on the Earth's resources'

"I believe that Contraction & Convergence offers us the framework for the only realistic way out of our current dilemmas if we are to do so in a planned and staged manner. With the governments of powerful countries and international organisations refusing to give it proper consideration, what better way to increase the model's profile and to show international solidarity than for private citizens to pick up the ball and run with it themselves?"

—Jonathan Dawson, President, Global Ecovillage Network (GEN)

Acknowledgements

Many thanks to all the wonderful people in the loose assembly of The Converging World, Schumacher Institute, Social Change and Development, and Go Zero teams; to Aubrey Meyer and the Global Commons Institute, Jeffrey Newman and the Earth Charter Initiative, Tony Hodgson and the International Futures Forum; to SCAD, for portions of text from their website; and to John Pittard, Helen Ayrton, Carole Moule, David Friese-Greene and Denise Perrin who contributed the photographs.

Preface

by Ian Roderick

The Converging World is based on the principles of Contraction & Convergence; it proposes that we reduce consumption of resources, energy and the emission of greenhouse gases in the demanding world while we accept an increasing consumption in developing and deprived sectors. It is a vision to which all communities can strive – each of us using an equal share that, in total, the Earth can bear.

Life is a constantly changing web of connections and dependencies. Sometimes we can see great and long-lasting structures and symmetry in these webs as if they were well designed, while at other times webs appear as tangles that are forming and reforming by chance and necessity. However, there are some people who excel at spinning these webs; they have an inexplicable ability to bring people together and to make things happen.

The Converging World (TCW) is a UK charity formed with great ambitions backed by a spinner of webs – a connector. My co-author John Pontin is a modest man, and he will hate having attention drawn to his efforts; however, over the last few years he has inspired many people to get up and do something. His influence works by the strength of his enthusiasm, and his belief that there is no point in just moping over the problems of the world – or waiting for others to provide solutions when there are 10,000 things you could do instead.

We will tell the story of The Converging World. Its aim is to tackle climate change, poverty, social injustice, inequality, waste and all the other 'crises and evils' that are preventing societies from providing rich, meaningful lives to all individuals. Nothing too modest in those ambitions, you would say, but you might justifiably ask: how on earth and what, where and when can it be done?

The Converging World contains the simple belief that we can achieve a better world through personal responsibility, through local initiatives and community linking. Some might consider this naïve, but there is a strong perception that direct, personal action can make a difference: we do not need to wait for international bodies to draw up global plans and then produce top-down legislation and taxation to force us to change. Our proposition is that we can alter our lives so that economies, wealth and well-being converge across nations and cultures to a level that the Earth can support, and we don't need to do it in isolation. If we link our communities we will learn from practical experience and communicating what Convergence means, while we are celebrating and enhancing those differences that define who we are and that really matter.

In asking 'how, what, where and when' will TCW do anything, we did not add in the question why? This is because the framework is familiar; it is almost unnecessary to ask why. However, to reinforce the message: our complex natural and cultural systems face many crises. Apart from wars, political and religious tensions, terrorism and social conflicts, our ecosystems are under stress and they are threatened with systemic collapse from over-exploitation and pollution. Easily accessible fossil energy supplies are reaching their peak of supply while demand for them soars, and the world's population is likely to carry on growing until at least 2050.

One consequence of this entirely profligate (but understandable and lovable) human activity is the release of greenhouse gases – the quantity of which is immense and growing – with the consequence that disruptive climate change is a near certainty, and may happen abruptly. We are in the century of crises. However, the perception of these crises is still limited: we continue to live in denial.

Those who don't deny the crises find it difficult to imagine an easy way out of this mess. However, they do find it easy to imagine a difficult transition to a world focused on survival – a return to a dark age. It is the nature of this survival worldview that we are exploring, and looking for ways to cope with it. Our values, which govern how we behave, are closely coupled to the ideas that we use to shape and plan our future. In our work we are seeking the ideal of 'The Converging World' that will act as a new framework for survival, with justice and equity rather than the old framework of survival for the fittest – those best able to fight their way to the top in a harsh age. This is the 'why' of TCW.

We will try to define and explore Convergence in this Briefing, and so provide a collection of ideas as part of the answer to why TCW is necessary. Convergence is a complex concept involving carbon emissions, climate change and energy, cultural diversity, differing values, technological change, human rights, spiritual challenges, political power, social struggles and resistance. All these dimensions interact as increases in population and consumption are stretching the planet's capacity beyond its limits.

Convergence means reducing the ecological footprint of some while increasing that of others. This is a difficult idea for those of us with a high-demand lifestyle; it implies shrinking our consumption so that others may expand theirs, and shrinking to such an extent that the total impact is bearable by the ecological, economic and social systems we inhabit. This is a tough target physically and spiritually, but we must meet it while not blaming anyone for the predicament we are in.

Convergence means resistance and struggle to obtain equal access to health, education and welfare. It would be wonderful if actions to reach these goals came from the top through international agreements. However, this only seems to occur half-heartedly, with an eye to the opinion polls, after concerted pressure is applied. As we get on with things as individuals and groups, it is a long road, and we will need patience.

Convergence means creating a just society where people can live freely, fearlessly, and receive a fair reward for their contribution to their community. We must seek methods to reduce the barriers to freedom that prevent this equality – we need fortitude to know that we can do it.

We will describe our ambitions for how TCW, as a new charity, will put these ideas into action. The remarkable work of an Indian agency called Social Change and Development (SCAD) has inspired us. We have started this initiative by erecting wind turbines in India – a simple, elegant project that has multiplying benefits for Convergence that will lead the way towards many exciting ventures.

Foreword
by Cletus Babu

Loving wishes from Amali and myself.

I am delighted to be part of The Converging World. It is a noble initiative and I am happy that we in SCAD are contributing to it in a small way. It is going to have a great impact in linking different sections of humanity to promote harmony and peace and above all environmental protection for the future generation.

It formed a part of our exhibition for the visit by Indian President Abdul Kalam, and he took a great interest in renewable energy and power generation. We had a Model Wind Turbine made, which had power to light up the street-lights of the leprosy village to demonstrate that we were able to facilitate government provision of electricity!

I would like to express my gratitude to you all at Chew Magna for your wonderful hospitality and commitment to SCAD. I thoroughly enjoyed my visits and enjoyed seeing the renovated Old Mill where it all started.

I wish TCW all success in its initiatives, and we are proud to the extent that we are associated with TCW's activities.

With love, appreciation and gratitude,

Cletus

Chapter 1

What does Convergence mean?

To converge means to merge, unify, unite, approach, and gather together. There is a dynamic quality to this word; it's as if we can always get a little closer. The opposite of converge is diverge, which means to move or draw apart, to vary, to deviate and to extend in a different direction. It suggests separateness, boundaries, and maybe disagreements.

We can see convergence and divergence in nature with co-operating and competing organisms making up complex ecosystems, which exhibit both these types of process to maintain a fine balance. How often do we say that in our social lives 'we have to get a balance', and how often do we blame an over-emphasis on divergent processes as the cause of the ills of our society?

For example, the history of the West since the Renaissance has been driven by expansion and by economic growth, so much so that we now claim that our economic systems are necessarily locked into ever greater volumes of production and distribution, which produces an increasing gap between those who have and those who don't – this is both among and within nations. Divergent growth not only causes tensions and conflict, but it feeds on itself and has led to a crisis point in our overuse of resources and disregard for waste. The current focus on atmospheric pollution is a direct symptom of a multitude of divergent processes.

It is tempting to suggest that we impose some opposite, converging process whereby we curtail overall growth and introduce the idea of equal rights to equal shares of all resources. Unfortunately there appears to be no will to vote this in, and we have tragic examples of how imposing something like this gets perverted and ends in disaster. But it may happen this way, as Noam Chomsky says:

> "Suppose it was discovered tomorrow that the greenhouse effect has been way underestimated, and that the catastrophic effects are actually going to set in 10 years from now, and not 100 years from now, or something. Well, given the state of the popular movements we have today, we'd probably have

a fascist takeover – with everybody agreeing to it, because that would be the only method for survival that anyone could think of. I'd even agree to it, because there are no other alternatives around right now." – Chomsky, 2002

This frightening prospect of a self-imposed, fascist convergent mechanism would probably contain horrors on a vast scale. However, Chomsky continues by saying that we don't have to wait for the disasters to happen; we have to create the groundwork to adapt and cope. These mechanisms, if we take a moral stand, should be convergent. They should bring people and people together, and people and nature together.

Balance is needed, and we need to hold and live with contradictions and even paradoxes. Convergence and divergence are both required within complex systems; they are opposites, but we need to find a way to accept the dynamic unity of these opposites that need each other. Communities do develop separately: they may pull away and diverge, but forces for convergence are required to prevent them springing too far apart; they need reasons and goals for uniting in common purpose for mutual benefit. It is our contention that our present world systems have too few converging or correcting mechanisms to do this. As a result we have extraordinary wealth in some parts of the world and abject poverty elsewhere, and this divergence feeds on itself.

The challenge of The Converging World is to discover and implement mechanisms for convergence in order to counteract the forces that separate.

At the core of any construct of a converging world is the natural world. We now have a perspective on this planet that shows how remarkable life is, and how life is essential to the maintenance of the atmosphere, the seas and even the rocks. Our actions are destroying this fine balance. We have introduced new, divergent processes by burning fossil fuels and peat bogs, and we have inhibited convergent processes like the absorptive action of the rainforests, by chopping them down and burning them.

Proper regard for convergent processes would restore that balance. It would see human activity as part of the natural systems: our waste will be Earth's resource, our resource will be Earth's bounty in mutual beneficial cycles rather than the destructive, one-way processes that we have currently created. All societies need to converge to similar benign relationships with the Earth. The model for those relationships is what we propose is at the heart of our future work, especially our relationship with the atmosphere.

The Converging World is a big idea. It derives in part from the 'Contraction & Convergence' principle proposed by Aubrey Meyer of the

Global Commons Institute (Schumacher Briefing No. 5), which sees, across the world, an equal per capita right to emit 'carbon'.

Contraction & Convergence

"Should Contraction & Convergence fail to occur, the globally asymmetric growth of emissions and income will continue, triggering rates of damage that will collapse security and overwhelm the [global] economy. The need to avoid this is obvious and absolute." – Global Commons Institute

The year 2006 may be a turning point in history. The media created a storm over climate change, and the last vestige of any dissent about global warming being real and serious almost disappeared under a hail of articles, publications, films and conferences and important reports.

A cynical view is that this media attention is transient; global warming fatigue will set in, just like fatigue over famine disasters or HIV/AIDS in Africa. Perhaps, though, we have attained some critical mass in the thoughts of enough people, perhaps some paradigm shift is under way and we, as yet, have no idea what the consequences of that may be.

If so, then major credit goes to the work of Aubrey Meyer and the Global Commons Institute (GCI), who have consistently presented and represented the concept of Contraction & Convergence.

Contraction & Convergence (C&C) is a framework for international action; it is generally considered fair and simple. C&C says that the right to emit carbon dioxide is a human right that should be allocated on an equal basis to all humankind.

Carbon dioxide is a useful proxy for all greenhouse gases, the concentrations of which are now threatening to produce irreversible climate change with potentially devastating effects on all life. The C&C idea is that we should estimate what levels of global emissions are acceptable to the Earth, which will then determine how much we need to cut back (contract). Once we know this, an equal per capita allowance is calculated which is used as a starting point for reduction and for trading (Convergence).

So by deciding an acceptable level for CO_2 in the Earth's atmosphere we can determine the maximum global emissions per year. This gives us an equal per capita amount: each individual's allocation. A full implementation of C&C would require global agreements and systems of trading so that the process of Convergence would occur by those countries unable or unwilling to reduce emissions paying others who were not fully utilising their allowance.

Contraction & Convergence was one of the major influences on the development of the Kyoto Protocol, which came into force in February 2005 as a mechanism to cut emissions of greenhouse gases. However, the Kyoto Protocol is an inadequate and limited response to climate change as, for example, the United States and Australia did not ratify it. Also India and China, who have ratified the protocol, are not required to reduce carbon emissions. These countries have formed, with Japan and South Korea, the Asia-Pacific Partnership on Clean Development and Climate to promote the development and transfer of clean technologies amongst themselves.

In summary, you could say that one half of the world is making a feeble attempt to reduce demand for fossil fuels while the other half seeks to clean up the supply. Nowhere is anyone implementing the concept of equal shares for all.

Contraction & Convergence remains as an ideal. It is on the table as a direct challenge to intergovernmental negotiations and to national policy makers. It has met considerable resistance, which suggests that it challenges the existing order that has got us into the mess we are in.

The Converging World is an experiment, an action story. Carbon, climate change, and poverty are its central players. In contrast to schemes which require governments to legislate and to force things to happen, our experiment starts with what individuals can do, and it makes the local connections that are necessary for changes in behaviour and for directly addressing the inequity that persists in our polarised world.

Globalisation and sustainability

As if carbon emissions and climate change were not enough, there are many other pressing global issues. We face the imminent prospect of the demand for oil, gas and water exceeding the capacity for supply.

A recent advertisement in *The Economist*, ironically run by Chevron, states that it has taken 200 years to use the first trillion barrels of oil and it will take only 30 years to use the second trillion [and there isn't a third!]. Throughout the world, water extraction from aquifers is increasing, and these fossil water supplies take tens of thousands of years to replenish – what happens to cities that run out of water?

This is all part of the context for Convergence. How do we allocate today the Earth's resources for the survival of a projected global population of nine to ten billion people in 2050 – and for their offspring indefinitely?

Minerals, energy, food and water are quite tangible. We also consume or control other less tangible items that are in some sense 'common'. Many aspects of knowledge, particularly scientific, medicinal, and agricultural, should morally belong to everyone.

The world has gone a long way to stating that everyone should have political and religious freedom and access to the law, health care and education. The preamble to the Universal Declaration of Human Rights says:

> "The recognition of the inherent dignity and of the equal and inalienable rights of all members of the human family is the foundation of freedom, justice and peace in the world."

The Declaration divides into two areas: political rights, and economic, social and cultural rights. So far implementation, in legislation, has concentrated on the political and freedom rights; little has happened on incorporating rights on employment, education, health and security. The Converging World embodies the Declaration of Human Rights in its philosophy; fully implemented, the declaration would produce a convergent world.

Globalisation and sustainability are two words that are over-used and have multiple definitions. They are shorthand labels for difficult, incomplete concepts. Are we concerned about the emergence of some form of sustainable globalisation, or are we interested in globalising sustainability? Who knows? We have to cope with contradictions: globalisation contains localisation, with both processes happening together and interdependently; sustainable systems comprise many unsustainable components, much like an ancient and seemingly sustainable rainforest which is only a collection of organisms that live and die within a boundary that we define.

The Converging World concept goes well beyond carbon emission reduction and trading, although this is a fundamental aspect requiring emergency attention. It is a vision of a world where everyone has a fair share of the resources that the Earth can easily provide without jeopardising its potential to support life in all its diversity. It is also a world where everyone has a fair and equal share of, and access to, human-created resources such as knowledge. The vision extends to an indiscriminate right to justice, health, education, welfare and security. In this worldview, environmental issues are inseparable from social justice.

Convergence means reducing the ecological footprint of some while increasing that of others. The ideal Convergence point is an equal, per capita footprint that, in aggregate, is somewhat less than the Earth can support: as a motto of The Converging World says, "99 percent planet living is the max."

In a sense, the tangible Convergence that we seek is a realisation of a convergence of values – material, economic, political and spiritual. It is our personal responsibility to reflect on what that means for our own behaviour. We are converging in diverse communities, each living according to its own local definition of sustainability but with a responsibility to the sustainability of all other communities.

So Convergence means many things. It is another shorthand word for the complexities of restoring a balance, where appropriate, between moving apart and coming together.

Our thesis is that our systems are deficient in convergent processes, that we witness too much divergence in many fields: social, economic, governance, well-being, academic, and spiritual. The values of the Western world, in particular individuality, self-improvement, freedom, liberalism and economic growth, have created a great civilisation but are divergent. Now these forces are coming up against the limits to growth on a finite planet and new, convergent forces are needed to subdue and correct them.

All such attempts at convergence are valid and necessary. If this is achieved by international bodies or national governments, then big changes may occur. However, TCW wishes to encourage those who just want to get on with it: by starting the process with the individual, with local community groups and with businesses.

Chapter 2

The Converging World:
the story so far

Convergence is action to achieve social, economic and environmental justice.

We will explore many of the theoretical ideas behind Convergence later, but one concept stands out: Convergence is a process. It does not define a state that we wish to attain; rather it shapes and evaluates the actions we take. We are going to take a pragmatic approach – ideas and reflections suggest actions, and then we learn from doing.

We will tell the story of how two communities are beginning a process of linking with Convergence as a central principle. It is a process that is deliberately slow and careful. Although there are few visible results, the importance of building relationships is revealed. Hopes and aspirations are explored and critically examined, and we explore the dangers of raising expectations that cannot be met. We suggest practical ways to learn about the process.

The story starts with an organisation working for improvement and sustainable development in southern India and how it got linked to a zero-waste project in the village of Chew Magna, near Bristol, in the UK.

Social Change & Development (SCAD) is an organisation that works in Tamil Nadu. It has been supported by the UK charity Salt of the Earth for many years. Cletus Babu, who runs SCAD, visited Chew Magna in April 2005 and this was the starting point for a tiny Converging World Group within a community-based project called Go Zero. This group was formed to raise awareness of how we should address issues of poverty and justice in the world if we hope to reach a zero-waste society.

One token aspiration was to match the village of Chew Magna (population around 1,000) with some of the SCAD communities in Tamil Nadu which have a population of around 5,000 – the ratio roughly equivalent to that between the one billion people in the developed North and the five billion in the developing South.

The SCAD story

SCAD was set up by Cletus Babu in 1985 to enable the people of rural India to stand alone, both socially and economically. It is now a thriving NGO, which works in the areas of education, health, community, agriculture and animal husbandry and supports more than 300,000 people in more than 450 villages. What it has done is produce a model, in the Gandhian tradition of social enrichment, which is now shared across India and the world.

Cletus gave up a comfortable life in the priesthood to make a difference to the quality of life of the rural poor. With two friends he took the first steps to lift people out of grinding poverty. They started simply by listening to the experiences of village people, and they slowly came to realise the communities' values and the problems they faced.

By gaining the trust of the people with whom they worked, and by taking practical actions, improvements slowly came. They had limited funds, but success appeared through encouraging villagers to support each other and to accept independence. Communities developed cohesion, and soon greater value was being placed on health and education – convergent forces that create a virtuous circle. Without grand articulation of theory, Cletus and his friends naturally understood the power of participation.

In 1986, they started to receive support from Action Aid. SCAD formed around 36 villages and the volunteers grew to nine people. With love and total commitment they inspired and encouraged the villagers to help themselves. Like so many stories where the will is intense and love so great, the resources to do things appear from unexpected quarters. SCAD acquired land and built a centre, making lasting changes to the lives of those in this southern part of Tamil Nadu.

In 1988 Cletus married Amali, who was working with tribal people and the slum children in Chennai. They are a powerful team. In 1989 they started the first SCAD school for rural children.

Indian President Abdul Kalam has met Cletus to discuss progress in rural development. In September 2006 President Kalam visited Tirunelveli to see the extent of the work and to meet members of the 'SCAD family'.

The president's vision for the future of India is set out in his 'Vision 2020'. He dreams of an India that is holding her own on the world stage in industry, commerce and technological developments, and empowering rural people so that every citizen is living above the poverty line, with good opportunities to earn a living, and access to education, health care, clean water, sanitation and electricity.

The vision of SCAD

The vision of SCAD has always been to 'reach the unreached', and with every year that passes, it grows from strength to strength.

Our mission at SCAD is to empower the downtrodden and under-privileged rural communities by uniting them and making them architects of their own development.

This can only be achieved by encouraging them to look at the root causes of their problems, and take a holistic approach to development.

Once we find a model of development that works, we tell the Government about it so that many more people can benefit from the work of SCAD, not just in the state of Tamil Nadu, but also across India and other continents.

The values of SCAD are summed up in this poem:

Go to the people, love them,
Live among them, learn from them,
Start with what they know,
Build on what they have.

But of the best leaders,
When their task is accomplished, their work is done,
The people all say,
"We have done it ourselves."
(Lao Tsu)

Life in rural Tamil Nadu

Tamil Nadu has many areas of extreme poverty; the main sectors of employ-ment are agriculture, salt production and fishing. The weather varies widely from drought to ferocious winds to monsoon rain. Work is seasonal and inconsistent, which inevitably means that income is unpredictable.

The tsunami of Boxing Day 2004 caused devastation in the coastal areas. Many people lost lives and livelihoods, shelter and employment. SCAD is one of many agencies that are working hard to repair the damage. These conditions are ripe for exploitation, since the poor have often borrowed from moneylenders and are now carrying the burden of large debts, sometimes inherited from their parents. It is common to find bonded labour, and children working long hours just to contain debt, let alone pay it off. To escape this misery, men often migrate to the cities. Families are left behind, and as a result they become fragmented.

The rural women of Tamil Nadu are particularly disadvantaged. They are dominated by men, who control every aspect of a woman's life. Women become powerless and completely dependent. They work day and night, walk miles for water, eat the leftovers of the family and take beatings or encounter 'accidents' when their families cannot afford to pay their life-long dowry.

Like all regions that rely on agriculture, the weather is something with which you live and to which you adapt. Monsoon climates are capricious. Some years too much rain falls too quickly, while other years no rain falls and the entire investment is lost, leading to hunger and malnourishment.

Villages rely on rainwater, but they lack the means to harvest what falls and there is little infrastructure for supply. Women are forced to walk for hours to carry water, or it is obtained from boreholes that are often brackish. With inadequate supply and poor sanitation, waterborne diseases are rife.

Most houses are usually small, thatched with coconut and palm leaves, with baked brick and mud for walls. They lack toilets, water supply and electricity, and cooking is often over wood or dung fires.

Millions of children in India do not complete school, and poverty is at the root of all reasons why this happens. Although parents dream of an education for their children, they need them to help out financially by working or by caring for younger siblings. Education is a luxury.

Access to medical treatment is difficult. The roads are poor, and bicycles and carts are often the only form of transport. Primary Health Centres often are many miles away, so communities have to rely on their own resources and traditional methods. Mortality rates are high, and the fabric of the villages can be threatened.

Whenever people live this precariously, the chances of falling into vicious downward spirals are high. Struggling hard to earn a living means lack of time for maintenance of infrastructure, and then illness lowers productivity further. Malnutrition and debt exacerbate the problems, which then demand higher and higher levels of work just to keep going.

What does SCAD do?

When SCAD starts to work with a village, the main objective is to get the villagers to a point where they can help themselves, as it is important that they should not become dependent on SCAD in the long term.

A volunteer from each village oversees all the development work, with skilled support from the field staff. This volunteer acts as a bridge between SCAD and the community.

Ownership and organising

A key factor in success is making sure that the community 'owns' the work. So committees of villagers called People Groups are formed:

> "If there is an issue that needs dealing with, who better to decide what should be done but the people who have the local knowledge, practical experience and biggest investment in making sure that it is successful?"

People Groups have overall responsibility to implement activities, but smaller groups are established to provide extra help and commitment, making sure activities in certain important areas like water, sanitation, village health and education, are driven through.

Women's groups

One of the most impressive aspects of the work is the women's self-help groups, which strive for the equal status of women, as well as raising awareness of gender relations and areas of inequity. These women's collectives initiate many of the social developments that lead to economic self-reliance. It has not been easy:

> "Many of the women have amazing success stories, facing imprisonment in order to stop large manufacturing companies stealing the sand from their rivers and avoiding money lenders, but one common theme amongst all is that their life has greatly improved from the one they can remember, where they were unable to speak in public unless through a male relative, and had not even travelled to the next village to meet their neighbours and see how they lived."

Health care

Breaking the vicious downward spirals starts with basic health care – this is SCAD's top priority.

The health team is made up of qualified doctors, nurses, health activists and experts in indigenous medicine. The approach is holistic, to make sure that all risks to health are considered and disease is reduced as quickly as possible.

Next to tackle is the cause of so much ill health. SCAD works hard to secure safe drinking water through the introduction of rainwater harvesting tanks and the deepening and renovation of traditional drinking ponds. Training is given, particularly to children and women, on safe practices to help prevent the contraction of waterborne diseases such as diarrhoea. Sanitation is a major concern, especially where there is a lack of water, and the risks around childbirth receive particular attention.

SCAD is working with the women's self-help groups and the village health committees to raise awareness about the importance of a good balanced diet. Through training and the provision of seedlings, SCAD encourages kitchen gardens, and women are taught how to cook better quality meals. SCAD has even created a tasty biscuit that has all the nutritious supplements required by children. Thousands of children look forward to the arrival of the biscuit van each week.

Education

SCAD has worked with the government to make the village schools more attractive, stimulating places where children will want to go. It has provided for the needs of pre-school children and those with special needs such as physical disabilities and learning difficulties. SCAD has established schools for child labourers, and the children of gypsies, the salt pan workers and those who are simply rejected from mainstream society. SCAD campaigns hard to emphasise a child's rights to education, and the importance that education will have on opening up future employment opportunities and what it will mean for the family and the community as a whole.

On the land

SCAD helps villagers to renovate their irrigation tanks, construct check damns and dig farm ponds so that they are able to make the best use of the rain when it does fall.

Farmers receive holistic training, which is tailored to their particular land. This involves live demonstrations that encourage the use of new and improved methods.

SCAD's tree-planting programme involves a wide range of trees, including bio-fuels, fruit trees and drought-tolerant species that are indigenous to India. This improves biodiversity, and will help farmers and families to have a healthy diet as well as providing additional income.

Life in the villages is perilous, but traditionally the ownership of livestock provides a supplementary income. However, with limited funds, it is difficult for many to do this, and if funds are available they are usually only enough to get poor quality livestock. SCAD provides good livestock to suitable and committed families with the agreement that they will purchase additional animals with the income generated by the first.

Vocations and new industry

SCAD gives vocational skills training to the rural villages to help introduce new and alternative ways of earning a living to the communities. This gives those who previously did not work a purpose in life and self-confidence.

Training in pottery, tailoring, fast food production, and outboard engine mechanics are just a few of the initiatives which have helped to bring meaning back to people's lives.

SCAD provides seed capital to many who are unable to take up conventional employment opportunities such as widows and those with learning difficulties. This enables them to become self-employed so that they can control their own finances and free themselves from the clutches of moneylenders.

The rural people have found tremendous strength in coming together as a group and reinvigorating their community values. People who previously felt alone are now part of a collective group who share the same problems and the same vision for development.

SCAD's objective is to withdraw slowly from the villages that they help and to move on to others that have not moved so far down the road to sustainable community life. For all agents of change, real success is when you are redundant.

Although huge changes have taken place, there is much to do and much to struggle for.

"Together they can make their voices heard and make the changes to their lives that they had always dreamed of but felt unable to do."

The Go Zero story

Let's now journey 5,000 miles away from rural India and back to the comfortable life of early 21st-century UK.

The converging world idea arose in the village where we both live. However, the story also has some origins within the Royal Society of Arts of which John is an active and senior member.

The Royal Society for the Encouragement of Arts, Manufactures and Commerce – commonly known as the RSA – was founded in 1754 by William Shipley, a painter and social activist, on a manifesto "to embolden enterprise, enlarge science, refine arts, improve our manufactures and extend our commerce".

The RSA started in the coffee-house culture of 18th-century London. Men of importance and influence would meet to discuss contemporary issues, and they would formulate plans to promote the arts, commerce and social needs. To celebrate 250 years of such discussions and influence, the RSA produced a new manifesto for the 21st century. This consists of five coffee-house challenges to shape its future work:

- Encouraging Enterprise

- Moving Towards a Zero-Waste Society

- Fostering Resilient Communities

- Developing a Capable Population

- Advancing Global Citizenship

As John was on the Advisory Council of the RSA for 'Moving Towards a Zero-Waste Society', it was not surprising that the Wales and West of England branch took up this particular coffee-house challenge. The concept was to gather interested people in a coffee-house to recreate the atmosphere of informal discussion that might lead to action. The branch held a series of meetings in Bristol sponsored by Starbucks Coffee!

At roughly the same time, the villagers of Chew Magna were collating the results of a survey for a parish plan. Everyone had been asked for their opinions on aspects of their environment and social organisation, and how they would like to see the parish develop. The outcome of this consultation was a report and an exhibition in the village hall.

John went to the exhibition and was soon in discussion with Marie-Louise, the parish clerk. It wasn't long before they were talking about moving Chew Magna towards zero waste – and why not? It was as if the parish plan came from a 'business as usual' worldview, while zero waste was a different perspective, one that accepted the need for fundamental shifts in thinking.

The next step from that simple discussion was obvious to John – it was to start a series of coffee-house conversations in the village, with no pre-conceptions as to any outcome or actions. Luckily Chew Magna has a great coffee shop with a very tolerant proprietor. John and Marie-Louise just asked around to find out who was interested and within days a group had formed. After these initial chats over coffee, several more gatherings happened – a few in the pub for those not around in the daytime. Each time the discussion was lively and soon the numbers swelled.

This first response was surprising – something had tapped deep into the consciousness and released energy. Not only were people concerned, even anxious, but it was apparent that they were keen to make something happen. When more than 40 people turned up one evening, it was too late to stop and four groups formed to start doing things.

Three of the groups were obvious clusters of interest. The Energy and Transport group had ideas ranging from the easy ones of efficiency (like insulation advice) through to ambitious schemes for micro-generation installations. The Waste and Recycling group formed to tackle the output streams (the rubbish), while the People and Consumption group dealt with lifestyles and input streams (goods and food).

The fourth group was slightly different – outward rather than inward looking; it emerged as people responded to a global / local dialogue. We were trying to do something about a global problem by our local actions, which meant taking responsibility for the effects we indirectly make on people and places far away.

One immediate and again obvious concern was sustainable travel: how much we travel and what we do when we are abroad. Another concern was about production of the goods and services that we import and the extended supply chains of which we have little knowledge. The local church was already promoting Fairtrade, but this was an area that we could support. The village also had long-standing links to Zambia and to Ghana, where a locally formed charity called The Atiamah Trust supported an HIV/AIDS orphanage.

The people particularly interested in these global issues held coffee-house conversations and formed into what they called the Converging World Group.

Open Day

The four groups developed at an early stage, and they created an urgency to do something. What we felt we needed to discover was the extent of the concern in the community as a whole. Were we the only ones interested, or were there many more who would like to get involved? We decided to organise an Open Day, and each group was assigned a corner of the church hall to display ideas and to host discussions. The children had a special area for making useful things from waste.

We delivered a glossy four-sided newsletter to every house in the area. This presented the ideas, and it invited people to come along on the day and to lend their support. It was with a degree of trepidation that we opened the doors – would anyone come, would they join?

It exceeded expectations by far: it was buzzing all day as if we had released the desire to talk about these issues with a view to doing something about them. We also learnt that there is a great deal of knowledge, skills and experience in a community like this. As a result of that day we had 150 people signed up, and thanks to Sally a name and logo for what we were creating: Go Zero.

The work continues with more newsletters, open days, surveys and meetings. Go Zero is now a non-profit company, with a range of projects aimed at awareness, monitoring, education and demonstration. In its small way it has stimulated and inspired other communities, but it is now in for the long haul – 20 years at least.

How did The Converging World emerge?

Early on in Go Zero's life John had introduced SCAD to us as a potential community project with which we could develop links. Cletus Babu had come to meet us even before the first Go Zero Open Day – while he was on his tour of the UK organised by Salt of the Earth. It seemed an ideal partnership. However, some people in Go Zero were sensitive to the possibility that this might detract from the efforts to support the existing community links in the village, especially if we were to try to raise funds for SCAD.

Go Zero generated considerable local press coverage that then built up to national papers, magazines and television. This exposure meant that other communities started to form, and we began to envision thousands of Go Zeroes across the country.

A dilemma arose. Go Zero was (and is) a village project focused in a specific locality, yet there was an ambition to spread it, to connect to other communities, to organise and develop a network.

So two threads to The Converging World (TCW) emerged: one is the desire to link communities across the continents, and the other is a vision of spreading the idea of community action and forming connections among the groups that spring up. Neither of these threads is straightforward.

A third thread in the development of TCW arrived with Go Zero's Carbon Club. This was formed to allow members to offset their carbon emissions.

We had had an evening meeting about climate change and offsetting, where the idea was presented that people could pay into a well-known scheme to get trees planted to absorb the approximately 10 tonnes of carbon dioxide emissions for which they are each responsible. This went down like the proverbial lead balloon. Go Zeroites weren't that naïve. They had gathered into a concerned group to do things. They were not interested in a scheme that asked them to pay up, wash their hands and carry on as before, however worthy the projects were on which the money would be spent.

This mismatch of visions was keenly debated. There was a kernel of something important in offsetting – it provided an accumulation of money which enabled action – but the feeling was that it had to be linked to carbon reduction in the here and now; it needed to work for changes with our own lives.

What emerged was the idea to start our own carbon offset scheme. To keep it simple we formed The Go Zero Carbon Club and invited members to donate on a self-assessed basis. By collecting this money ourselves we had it under our own control as long as it was within the stated objective of our constitution:

"The club will apply all monies raised to schemes that aim to reduce greenhouse gas emissions or otherwise mitigate the effects of climate change, or counteract environmental and social degradation."

This third thread of a locally controlled carbon-offsetting scheme was one more component in the tapestry that led to The Converging World. At the same time, the RSA was developing concepts for carbon trading and promoting the idea for personal carbon allowances. We can see that many ideas were beginning to surface.

John really took to the idea of linking with SCAD, and the group in Go Zero was a prompt for his efforts. However, he is an inveterate connector so he organised a meeting in Bristol of an eclectic group of concerned individuals and young people to explore what might be achieved with SCAD. The result of this meeting was a challenge paper, which contained a short statement of a vision:

> "Cutting the carbon is an emergency issue and is central to this challenge. A full implementation by governments of the Contraction & Convergence concept is not happening – we have piecemeal attempts like the Kyoto protocol. We want to explore, through linking communities, what mechanisms might allow individuals and communities to 'trade' their per capita allowances directly – locality to locality.
>
> "As this process happens we will explore complementary convergence, again trading our allowances. These are actions that build on this practical linkage. Some possible areas are:
>
> • Trade: suitable products that can be offered either way.
>
> • Tourism: opportunities for exchange visits and hospitality.
>
> • Education: training for basic skills, opportunities to learn from each other or learn together – with modern communication technology much can be done without the need for travel.
>
> • Enterprise: looking for ways to increase employment, for new business opportunities.
>
> • Arts and entertainment: producing films, story books and similar products.
>
> Although many of these ideas are just dreams, it is only through dreams that we can glimpse at possibilities."

Following the Bristol meeting, which was set within the context of Aubrey Meyer's principles of Contraction & Convergence, the next step was to explore ways to establish mutually beneficial relationships and projects between Chew Magna, the West of England and the UK in the 'developed' world, and two communities in Tirunelveli District, Tamil Nadu, India in the 'developing' world. The goal was to use and build upon long-term relationships between these villages to illustrate how Contraction & Convergence principles can be implemented, proven in practice, and adapted on a local scale to illustrate global possibilities – once again, nothing too ambitious!

John started plans to visit India. However, one young person at the Bristol meeting was Bremley Lyngdoh, a co-founder of Global Youth Action Network (GYAN) that works to implement the Millennium Development Goals with governments, the private sector and other NGOs globally. Bremley was about to go to India, so John asked him if he could extend his trip to visit SCAD.

The goal was to develop links between the participating communities both in the UK and in India. It was considered important to understand what kind of support the people from Tirunelveli might need from the people of Chew Magna, to identify what role each community would play in this partnership, and find out what kind of exchange of ideas and initiatives would generate 'win-win' outcomes.

Bremley arrived back in the UK with a proposal from Cletus and his team building on the idea. Would it be possible to secure a windmill that would generate clean energy in order to power many villages?

The threads of many ideas were drawing together as John went to India to see SCAD at the beginning of 2006. It was on this trip that the concept for wind turbines seriously came into play. The idea came first from Bristol property developer Mark Tucker, a keen supporter of sustainability initiatives.

A vision of linking

Seeing the wind turbines in India was probably the turning point. The visions of linking could be realised in a 'triple win'.

We could help to meet the growing demand in India for electricity by supplying renewable energy that substitutes for fossil-fuel-generated power. The carbon emissions saved in this way are increasingly valuable to businesses and individuals in the UK who wish to meet their obligations to reduce but, as yet, can only partly achieve that by changing what they do.

And so a triple win appears: one, a reduction in carbon emissions; two, income from selling electricity is channelled to poor communities to achieve sustainable livelihoods; and three, donations, in exchange for saved carbon emissions, are used to fund projects and campaigns to contract consumption in rich communities.

Of course this would also bring added benefits. We will achieve even bigger wins than these three put together by forging links between communities, by changing hearts and minds and by giving hope, through action, that we can meet whatever crises unfold this century.

An elegant strategy

"As a child I understood how to give. I have forgotten this grace since I have become civilised." – North American Indian saying

After many discussions and many thoughtful hours, the idea for The Converging World emerged. Its immediate aims were simple.

We will raise donations from wealthy individuals and businesses anywhere in the world so that we can install renewable energy facilities like the large 1.5 megawatt wind turbines in Tamil Nadu, southern India. These installations will save carbon emissions that would otherwise be created by burning coal (about 60 percent of India's electric power comes from coal).

We will sell the electricity in India onto their grid. Some of this income is used to pay off loans, and then it will buy more turbines to multiply our effect, but we commit 25 percent of the net proceeds to SCAD. SCAD has worked tirelessly for sustainable livelihoods in this region for more than 20 years, but there is a great need for more money to bring thousands of people out of desperate poverty.

Meanwhile, back in the West, our 'green' electricity is offered as a way to counterbalance our over-consumption of fossil fuels and unfair use of natural resources. The Converging World is a membership organisation. Any individual, business or school may join if they are prepared to work towards Convergence. Members are invited to make donations that reflect their CO_2 emissions and their impact on the planet.

Through suitable pricing we set these donations against the carbon emission savings we are making in India. These donations are used to address the underlying cause of the climate change crisis. They fund projects that reduce emissions: projects such as community-scale renewable energy installations, educational campaigns for energy efficiency and against over-consumption, and social programmes like reducing crime.

It has been described as an elegant strategy, and Aubrey Meyer has called it a beautiful model. It has also been called a triple-decker sandwich approach, as it has three distinct benefits.

Young children from a SCAD village.

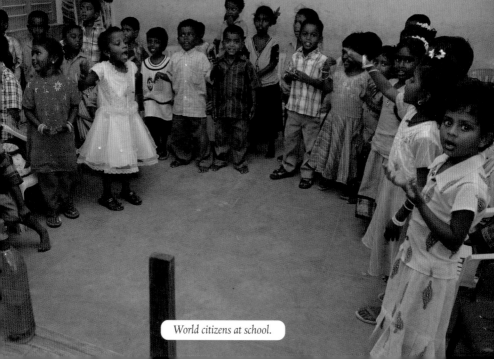

World citizens at school.

Life for some in rural Tamil Nadu.

The SCAD Polytechnic.

A visit to the Farm Science Centre.

Cletus Babu with John Pontin.

The Go Zero banner.

Go Zero promoting renewable energy at home.

Go Zero promoting fair trade.

Chew Valley School visit to Go Zero.

A Suzlon turbine under construction.

Children in a SCAD village with the energy to dance!

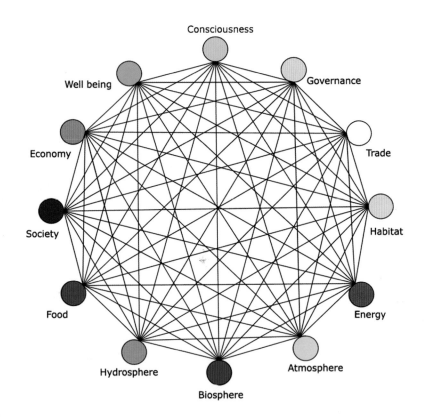

The 12-node World Model.

Reactions

The occasion of the Schumacher Lectures in October 2006 was an ideal opportunity to present The Converging World. It was a follow-up on the previous year, when John had first talked about SCAD. So this time Ian stood up for 15 minutes of fame. We also ran a seminar to allow people to tell us what they thought about the idea.

What reaction did we get from this discerning audience? For some, The Converging World seemed an exciting, ambitious and inspiring project. At the other end of the spectrum we had people who felt that we were betraying the 'movement' by jumping on a discredited carbon offset bandwagon. In between we had many confused people. This was a worrying and disappointing outcome.

We canvassed opinion from friends by sending out a two-page description of TCW. We received a measured and valuable response from Forum for the Future.

The good news was that we had a fantastic idea that was timely and engaging at many different levels – individuals, communities, businesses and government. It was 'joined up' in the way that it embraced local and global, North and South, personal and organisational. We were pragmatic by looking at the best use of resources and providing renewable energy in the South rather than in the UK. It was a creative idea that will appeal to many different people and opened up a host of opportunities. It was focused on measurable outcomes and, above all, it was intellectually robust, being derived from Contraction & Convergence.

The bad news was that it was not working well at the moment because it was too complex. There was a complexity of language, a complexity of structure and a complexity of ideas and concepts. It was far from being something you could explain in the time it takes for a journey in a lift.

The language problem was our fault. We knew what we meant, but we were used to talking as dyed-in-the-wool activists with our own jargon and style. We had not put any thought into communicating to people who had little idea of these issues – our enthusiasm ran ahead of us.

The complex structure, when you are aware of how TCW works, makes sense. However, instead of trying to simplify our description of the structure we had complicated it further in a belief that by introducing more layers and different organisations we were adding weight to our proposition. This was a mistake that prompted the comment:

"For an outsider who knows absolutely nothing about any of these organisations, this is a completely impossible labyrinth to negotiate."

The criticism that we had a complexity of ideas was a major problem. There were so many overlapping concepts buzzing around, so many sudden tangents heading off in different directions, so many different assertions and cost estimates that were obvious to us, but were anything but obvious to people not directly involved in the project. We needed to keep it simple.

We weren't marketing experts, but marketing was the skill required. We had to put ourselves in the place of potential supporters, donors and participants. We had to get the message right and present it in different ways to suit the different types of people we wanted to affect.

There were more problems, though. These weren't about presentation, they were more fundamental. The world had gone 'offset crazy' in recent months, and although The Converging World initiative clearly offered a lot more than 'just another offset scheme', that was still the territory in which it had to be sold.

To be competitive meant that we had to address this crunch issue: how do we get to be the best offset scheme going, how is it going to be possible to verify the schemes funded via The Converging World? Verification is a big part of what offset providers offer, and we had to play that game to be taken seriously.

Although The Converging World story had the carbon offset thread weaving in and out right from the initial Go Zero Carbon Club, we only slowly came to realise that it was such an important part of developing our work. We were seriously at risk if we weren't able to operate at the highest levels of integrity in this rather Wild West marketplace.

Carbon offsetting

The growing concern with climate change has led to an increasing acceptance of the need to reduce carbon emissions in the developed world. The aspiration is to be 'carbon neutral', firstly by reducing emissions as much as possible by efficiency methods or restructuring of operations, and then by offsetting, through a financial contribution, the emissions that cannot be eliminated.

These financial offset contributions are invested in schemes that prevent carbon emissions (e.g. renewable energy schemes substituting for fossil fuels) or they capture carbon from the air (e.g. planting trees that will absorb carbon dioxide). If these schemes are truly additional, then the net effect to a company or individual if they purchase sufficient offset credits is a carbon neutral position.

Organisations like the Carbon Neutral Company and Climate Care and many others have been established to help companies reduce and voluntarily offset their emissions. Equally, they can help individuals who wish to realise a carbon-neutral lifestyle.

How did this all start? A lot of this was modelled on the success of trading schemes to combat ozone depletion and also atmospheric sulphur pollution. These were the background models that helped to determine the Kyoto Protocol, which was developed by the United Nations Framework Convention on Climate Change (UNFCCC). Another major input to this process was the concept of Contraction & Convergence (Meyer, 2000). This is based on the principle that global emissions of greenhouse gases should be steadily reduced in the North as those in the South expand, eventually to achieve an equal global per capita level of emissions. Underlying this is the call for reformed global governance to make sure this is completed with equity and justice.

Under this system, developed countries that had exceeded their allowance could purchase unused allowances from developing countries and so a system of allowance trading could be established. Developing countries would have access to development funds by right instead of through aid and developed countries would have a strong incentive to reduce their emissions.

Since Kyoto, regulated carbon trading has become established, most notably utilising the Clean Development Mechanism (CDM) which yield Certified Emissions Reductions (CER) carbon credits (tonnes of CO_2 equivalent saved).

One of the best known carbon offsetting schemes, called the Emissions Trading Scheme (ETS), was introduced by the European Union in 2005 and is obligatory for certain manufacturers, utilities and other heavy industries in all EU countries. This sets individual caps or allocations on carbon emissions for these companies. If they pollute more than this, they have to balance their books by buying in carbon credits from others who have polluted less than their allocations require, or they can buy credits from Clean Development Mechanism projects that are usually in developing countries. The scheme was, however, somewhat discredited when it emerged that most countries (excluding UK) had set allocations for industry which were actually above the companies' existing carbon emission levels. Nevertheless, this example illustrates how carbon reduction and offsetting are becoming business considerations, whether through a statutory requirement (as in this case) or as an element of corporate social responsibility. The EU ETS ends in 2012 when the Kyoto Protocol ceases, but it is expected to run after this within a reframed and extended successor to the protocol.

The voluntary carbon offset and retail markets

The voluntary market refers to companies, governments, NGOs, and individuals that purchase carbon credits or offsets for purposes other than meeting regulatory targets. This is outside any mandatory scheme.

If a company or a government organisation wants to offset their emissions they would first undertake an analysis of what impact they were making on the climate: that is, they would determine how many tonnes of carbon dioxide they emit each year. There are many consultancies and companies able to do this analysis, and at the same time they would identify energy efficiency measures to reduce emissions. A comprehensive review of energy use of both direct and indirect emissions could lead to significant savings or identify opportunities for new products and services. When this process is complete they have a number of options, but the main route is to purchase carbon credits from offset retailers.

The retail market refers to companies and organisations that invest in offset projects and then sell portions of the emission reductions with a mark-up. There are different levels of verification for offsets, but recently proposals are seeking to tighten up standards.

One big challenge for carbon offsetting is accountability and transparency. Early schemes have been criticised on a number of fronts: it is not always clear where the carbon offset payments are going; when this is examined, the money often appears to be administered badly; and there is no long-term tracking of carbon offset projects (for example, inappropriate planting of single-species trees can actually damage the local environment).

The idea that a business or individual can simply pay to offset the problem has also received considerable criticism. Many of the schemes are in developing countries, giving the impression that the West can continue with business as usual without any regard to the bigger picture.

To bring some order to the scene the UK government is to set standards for carbon offsetting schemes, which will have to issue certified credits and be assessed to international standards, particularly to meet the Clean Development Mechanism rules under the Kyoto Protocol.

Environment Secretary David Miliband has said: "Offsetting isn't the answer to climate change. The first step should always be to see how we can avoid and reduce emissions". However, he agreed that offsetting has a role because "some emissions can't or won't be avoided".

The Converging World proposes what it calls a second generation carbon offset scheme (or, to change the language, a carbon reduction fund). This attempts to address existing criticisms by committing to the following. The scheme:

- Primarily seeks reduction of carbon emissions in the 'demanding' world;

- Addresses social, economic and environmental injustice, particularly in developing countries;

- Promotes community development and connections between communities across the world for sustainable livelihoods and low-carbon living;

- Works with local agencies as partners seeking critiques on what we do;

- Campaigns for an end to the unsustainable consumption and unsustainable growth logic of our societies and a reformation of the carbon trading concept;

- Goes beyond carbon in two ways: firstly, it looks for positive impacts not just carbon neutral; and secondly it considers the rest of the ecological footprint.

- Commits to learning, with the ambition to conduct 10,000 experiments in Convergence.

Does a second generation scheme, like the one that TCW is organising, really answer all the criticisms?

Critical views

Many people are suspicious of carbon offsetting. It's a Western, imperialist, capitalist's ploy to pay their way out of trouble at the expense of the poor in the developing world – again. Worse than that, it could also cause the release of more greenhouse gases than it saves, as it disrupts economies and ecologies through ill-conceived projects. It is only 'greenwashing', that allows businesses and individuals to continue their existing fossil-fuel-hungry ways.

However, at its most basic level the economic equation seems to make sense. What are the objections to the carbon offsetting idea? Well there are quite a few.

Measurement

The first and simple problem is all about quantifying how much carbon dioxide is emitted, and whether we can know the real cost of saving any amount of emissions. These are technical issues, and over time we should be able to derive a usable estimate of what is emitted and the true cost of eliminating a tonne of carbon dioxide. It may not be 100 percent accurate, but then the whole issue of greenhouse gases and climate change is based on estimates of the behaviour of a complex system. This is an objection to the implementation, not to the principle of offsetting.

Unintended consequences

A related objection is that we might not know the full consequences of the actions that we take to eliminate emissions somewhere else. Planting trees sounds good, but if the land is first stripped of full-grown oaks that are then burnt as fuel and little saplings planted in their place, something doesn't feel quite right. In any case, tree planting is a controversial method of offsetting as the trees will one day die and release the carbon back into the atmosphere – it is only a holding game. If done unwisely, it can be a disaster to an ecosystem by adversely affecting soil quality, the water table and the biodiversity of the area. The problems go on. Do these schemes displace indigenous people in the scrabble to spend the money?

Haves and Have-nots

> "This corporate offset culture magnifies inequalities between the haves and have-nots as the South becomes the carbon dump for the over-consuming North. The threat to indigenous peoples and peasant communities is especially severe, as destruction and/or loss of access to forests for many peoples would destroy their livelihood. The First International Forum of Indigenous Peoples on Climate Change stated 'sinks in the CDM would constitute a worldwide strategy for expropriating our lands.' "— Carbon Trade Watch

'Unintended consequences', as an objection to carbon offsets, is valid, but in time we will learn and get better – again this is not an objection to the principle of offsetting, just to how we do it. Responsible offset companies will continue to offer tree planting where the evidence is secure that it is a wise use of the funds, especially when it is coupled with the sustainable development of local communities. Efforts by organisations like Carbon

Trade Watch and CDM Watch are vital to expose bad practice.

Installing renewable energy facilities in the developing world is an alternative that appears to get around many of the criticisms of tree planting or of handing out low-energy light bulbs in poor neighbourhoods. Renewable energy generation is an easier way to account for emissions. A wind turbine, for example, will generate electricity that is fed onto a national grid and used to boil a kettle, light the streets or run TVs. Consumers are supplied with power that would otherwise be created by a fossil-fuel-generating station probably burning coal. Of course, the real gain comes when the quantity of renewable electricity is sufficient and stable enough that whole power stations are withdrawn from service or not built in the first place.

Renewable energy is an engineer's paradise, since they can cost the wind turbine easily and measure the power produced accurately. Elsewhere scientists have solid estimates of the amount of carbon dioxide and other gases that coal-fired stations emit for each kilowatt-hour of electricity generated. A clear measurement of carbon emissions saved is produced that includes a deduction for the carbon emitted in the manufacture of the turbine, preparing the land and installing it.

The offset company now has an accurate cost per tonne of carbon saved, which it can offer to those who wish to become carbon neutral – an extra is added for the company's operations and if it is profit-making then the shareholders should get a dividend as well.

What about the unintended consequences of renewable energy offset schemes? There are obvious issues to guard against, like making sure that local people are not displaced from their land, or otherwise disadvantaged by where the turbines are located. What are the credentials of the companies operating the wind farms? Are we happy that all the stakeholders are happy and no obvious exploitation has occurred? But these are objections to implementation, to how we do it.

No substitute

A major argument against renewable energy offset schemes is that the energy is not a substitute for fossil-fuel-generated electricity. Instead it just adds more capacity which increases consumption leading to greater economic growth. Indeed this is sure to happen in a market economy by the law of supply and demand. As more electricity is available the price goes down, and so demand goes up and more is used. Countries like India have a seemingly insatiable demand for power: their programmes to develop their generating industry are in place for years ahead; and renewable energy installations, the capital for which comes from the offset companies of the West, are an extra layer on top that will just accelerate the growth of the economy. The extra supply will allow the grid to expand and more people will buy TV sets, kettles and washing machines, all of which will create huge amounts of carbon emissions in production, transportation, use and disposal. So the whole idea of offsetting this way is a fallacy.

This objection ignores the bigger and longer-term picture, which we will come to shortly. It is also a rather imperialist view, since it implies that these communities should be denied the benefits of technology as otherwise they will join us in a clamouring, demanding world that depletes resources, pollutes the environment and increases emissions – let's keep them corralled in the state they're in.

Moral hazard

However, the main objection to offsetting is the shift of attention away from what is called contraction. Because I have paid to offset, I have an excuse not to reduce my carbon emissions in the first place. This is a phenomenon that economists call 'moral hazard'; it is similar to what happens with insurance that produces contradictory, risky and wasteful behaviour.

It induces a feeling that you have covered the risks and can now live without worry. I've paid to offset my flights to India so I can now fly to America without thinking that I am adding more than my fair share. That nagging feeling that may make me question why I need to go to America is removed.

This moral hazard objection to carbon offset schemes is valid, and is only countered when we appreciate the full story and do not let up in our efforts to reduce our carbon emissions in our everyday lives.

It is a question of values. Being concerned primarily with paying off your guilt by carbon offsetting is a cop-out. If you consider the planet worth saving, if you really value handing on a decent and liveable future to your children, then carbon offsetting comes as the icing on top, but after you have done everything else you can to reduce your own emissions, and you are committed to a path of efficiency and reduction as opportunities appear.

Futility

Is it all just a waste of time and effort? Isn't it like rearranging the deckchairs on the Titanic? Nothing is going to stop the expansion of the economies of countries like Brazil, Russia, India and China. These expansions will dwarf, in months, the amount of carbon dioxide we might save. The UK is responsible for only two percent of the total – why bother?

Robin Oakley of Greenpeace states that:

> "So-called offsetting is better than doing nothing, but only just. It's like smoking 20 cigarettes then going for a run to feel less guilty. As long as British vehicles are pumping tens of millions of tons of CO_2 into the atmosphere every year, no amount of investment in clean energy projects built thousands of miles away will reduce the effect that our emissions are having on the climate."

This is a counsel of despair, and however understandable that may be for all of us at some time or another, it does not appeal to the spirit and determined delight with which we should face up to the immense challenge of climate change. As a nation the UK has an obligation to do so, and it has many debts to repay – it is still the fourth largest economy.

Perpetuating neoliberal, global capitalism

The arguments against carbon trading and offsetting are deep and complex. By effectively commodifying property rights in the Earth's carbon-cycling capacity, governments and businesses have claimed and absorbed mechanisms to address global warming into their existing market-based systems. The unspoken objective is to perpetuate those systems that are heavily dependent on fossil fuel energy.

With huge fanfares in the media, with large numbers of consultants, negotiators, international bodies, verifiers, traders and other riders on the carbon gravy train, the wagons roll, making any other approach to global warming, such as broad-based political organising, exceedingly difficult.

By occupying the ground and making carbon trading appear as the mainstream, any other idea is branded as alternative and therefore much more difficult. While attention is diverted to implementation, verification and the mechanics of carbon trading, governments can ignore political pressure exerted upwards from a participating public that would focus on:

- big public works – decentralisation of power generation, tidal schemes, public transport infrastructure and many others

- removing subsidies – making the market work by exposing the true costs of fuels and other fossil-fuel-based materials like fertilisers and plastics

- applying subsidies – stimulating the renewable energy market by better pricing

- green taxes – and other non-trading market mechanisms

- legislation – requiring cuts in emissions

- action to enforce legislation – making it work

- debt cancellation and the reform of world trade to liberate countries to move ahead rapidly with renewable energy.

In an almost subversive way. The Converging World will take back the ground. It is intent on using funds to stimulate community development, building a grassroots movement to address carbon reduction and exert pressure to achieve structural change in our society. It will achieve this by playing the voluntary carbon offset game, through clean and verified means, in partnership with those affected. Maybe we shouldn't tell people this?

Other criticisms

The criticisms of offsetting go on and on:

Money from offsetting is better spent on carbon reduction in the west directly; you could cut more carbon in the UK from effective behavioural change programmes targeting low-carbon lifestyle choices.

Offsetting is just spin, a way for companies to draw the wool over our eyes while they carry on as normal. All the better for them if the schemes are in the South so we can feel doubly good as we salve our consciences.

Carbon prices have been driven too low by competition among offset companies – the direction should be towards increasing the price.

Verifying offset projects is bureaucratic and costly – jobs for the boys again.

There is great confusion over standards and regulations – best to keep away and not put your reputation at risk.

If governments were to buy offsets, wouldn't this be a misuse of public funds?

Will governments switch development aid into offsetting schemes?

If offsetting becomes a major mechanism will it not crowd out any other scheme such as international cap and trade, or personal carbon allowances? Will we get locked in to offsetting which may prove suboptimal, or even harmful?

The real challenge is to keep the carbon in the ground:

"This is the carbon con. Offsets do little to challenge our consumption of fossil fuels. And if we are to avert the worst excesses of climate change, we must end our reliance on those fuels quickly. Offsets do not fundamentally challenge the huge inequities in the world. In fact, they sometimes make them worse. Offsetting doesn't pressure companies to switch from fossil fuels to renewables or encourage governments to regulate polluting companies. It doesn't stop airport runways being built, planes being flown, cars being driven or even coal-fired power plants being brought online. In fact, it encourages them to continue and expand. It feeds on the good intentions of consumers and ethical business so that the fossil-fuel industry can thrive." – Adam Ma'anit, *New Internationalist*, July 2006

We would like The Converging World to be a 'joined up' way to answer all of these critical points. The most difficult question is whether it would be better to abandon offsetting as a mechanism because we may get locked into it, and we may then be unable to adopt different approaches.

Benefits

In this critique so far we have concentrated on the disadvantages of offsetting, but what about other beneficial, unintended consequences?

It may seem a weak benefit, but debates over carbon offsetting have certainly raised awareness of climate change and pollution in general. It is possible that this will drive grassroots pressure for political change.

Good offset schemes with an integrated and holistic approach can have positive social benefits. Communities in the South may welcome renewable energy installations that stimulate local employment and the economy, or provide managed plantings of trees that restore land or even reclaim deserts.

If large companies adopt policies for carbon reduction and offsetting, they may well extend this down their supply chains and mandate that their suppliers adopt similar policies in recognition that the ecological footprint of a company is often outsourced. This influence expands awareness and action to reduce emissions.

Where do we go from here?

Let's turn to the question posed on the RSA's CarbonLimited website:

"So is offsetting a good idea or not? Is it just a way to salve our conscience that does more harm than good? Or can it reduce our emissions while helping to tackle the most serious development issues in the world?"

What are the criteria by which we can measure and value the different approaches that people are taking to tackle global warming? This is not a simple question to answer. One seemingly incontrovertible criterion would be to what extent any action accords with the principle of keeping coal, oil and gas in the ground. However, to apply that means exploring the full systemic nature of our society over a long period.

Maybe David Miliband is wrong: maybe offsetting should be the first thing we do rather than the last? Imagine if every company and every individual were obliged to pay the appropriate amount for their carbon emissions – truly internalising the costs that we dump on the natural world. That huge, annual sum would be used for carbon reduction, renewable energy, sequestration and all manner of schemes to 'cut the carbon'. Every company and individual would have a massive incentive to minimise their own emissions.

To achieve such obligations would require either an unprecedented political determination to impose it from above, which is most unlikely, or a moral understanding that comes from the grassroots, from communities and individuals that voluntarily decide to do it because it is the right thing to do. The Converging World follows that latter route.

You may still have doubts and uncertainties about carbon offsetting; in fact you should, as many commentators have made justifiable criticisms of the concept.

Building the charity

It was early in 2005 that the small community group formed in the village of Chew Magna. It called itself Go Zero, and set about changing people's behaviour towards sustainable living – it has a long way yet to go. Many other communities have started up with similar aims and are doing remarkable things. The Converging World has its origins in Go Zero and the links that were made with SCAD in southern India. But what will TCW actually do?

Let's start the story with planting wind turbines in India. We are approaching large businesses and wealthy individuals, anywhere in the world, to become donors so that we can purchase wind turbines near to where SCAD operate. To maximise our effect we are using these donations as down payments of around 30 percent of the full cost, and we are borrowing the rest of the money from a bank. These loans will be repaid out of some of the income from selling the electricity generated. The turbines will last for at least 20 years so we can forecast a considerable surplus of income in their lifetime. Right from the beginning we will direct some of this surplus income towards sustainable development projects with SCAD.

TCW has established an operating unit called Converging World Renewable Energy Ltd which will negotiate and manage the technical and legal requirements to make sure that the turbines work and that we can obtain verified and certified carbon credits. We have a developed an excellent relationship with companies like Suzlon, India's leading wind turbine manufacturer. They are delighted to see a scheme that is from the charitable sector and are co-operating with us in every way.

We are building our relationships with donors, and presenting an attractive picture to get their commitment to the idea. It requires a level of skill in presentation to show how the integrated nature of the project produces a multiplier effect so that donations will make a double or even triple difference. It is a compelling story for larger companies that need to be seen working on their social and environmental responsibilities, as it addresses global issues as well as communities – TCW expects to involve companies and their employees in local schemes.

We are also building our relationship with SCAD through schemes like a media centre, somewhere to start producing films and website content about life in rural India. Through this exchange of information we will develop the personal links essential to drive cultural convergence.

TCW has been fortunate to have the active involvement of film-maker David Friese-Greene. David has been involved with SCAD for some time – he made a beautiful film about life there and the work that SCAD did. His sensitivity to Indian culture has been invaluable, as we have set about building links with the villages in Tamil Nadu.

Membership

As the turbines come on stream the carbon credits become available back here in the UK and the rest of the 'demanding' world. We are inviting any business, large or small, and individuals to become members of TCW and in return for donations we will retire carbon credits – if requested. This allows individuals and businesses to claim that they are offsetting or even carbon-neutral. The language we are using is notable. We seek donations as a charity, yet if a business or an individual wants to offset then we will allocate credits to them by retiring them from any further use – we don't hand them over. In this sense a business or a person can 'buy' carbon credits and satisfy their need to offset their carbon emissions to whatever degree of accuracy they wish.

However, as a member of The Converging World we will expect you to adopt certain behaviours and attitudes, namely to reduce your consumption of energy and resources, to deal responsibly with your waste, to adopt an ethical approach to your activities both at home and at work, and to support efforts for justice and equality throughout the world.

Community groups and schools

Convergence is a process, a series of many small steps over a long period. These steps are easier to take and often most effective when you are part of a community that is working together to achieve change and doing more than individuals can do by themselves.

The money raised by membership donations is called the Sustainable Community Fund, which is directed to large-scale carbon reduction projects in the UK or elsewhere in the demanding world: projects like renewable energy installations or social change with a 'green' tinge. We would like many community groups, such as the Go Zeroes of this world, to become affiliated to TCW, and the fund will support them with resources, networking and the exchange of ideas. In return we would just ask that they advertise and promote us. Community groups will be invited to apply for grants for their own projects that may exceed their capacity to fund themselves – particularly renewable energy or waste reduction schemes that require large amounts of money.

We also invite schools and colleges to join and develop links with communities and other schools in countries like India.

TCW will act as a network for the exchange of ideas with regular bulletins, publications and films, and practical advice to help groups form, develop and extend their influence. This will create links between communities in different parts of the world.

Our work in Tamil Nadu is our first experiment, although quite a large one as we would like to establish at least 40 turbines. However, we will look for similar opportunities in other parts of the world. These projects will be locally controlled and will directly link communities across the world. Although the poorer parts of Africa come to mind it is equally possible that we will work with the poorer parts of the UK or other developed countries – inequality is a patchwork across the globe.

To build TCW will require people, money, resources and systems. We are blessed with many volunteers, but more are always welcome. We rely on the generosity of our well-wishers and supporters. We believe that the message of Convergence is compelling and that the activities of TCW will allow many communities to act now because we cannot afford to waste any time. It is an ambitious project.

Plans

What are our ambitions?

Climate change is a unifying problem – Convergence is a unifying principle to tackle it. A world war provides a common enemy but we hesitate to push this analogy as we do not wish to cast other people or nature as enemies. However, let us say that Convergence articulates a common mechanism to judge whether policies, plans and actions fit towards a common goal. Likening the situation to a war is only because of the scale and urgency of the problems we face.

Obviously we want The Converging World to happen, and in this case we might relax our view that small is beautiful! However, there are many other things that we want.

We want The Converging World to become a major force that will affect the material and spiritual behaviour of individuals everywhere in the world – bringing them closer. Our challenge is to develop the concept of Convergence with the full participation of all involved – this will require a lot of learning and understanding. We have established ideas already that will shape what actions we will take – these need to be challenged. However, action is our way to find out and reflection on action is how we learn.

We want to create many carbon reduction programmes. These we call our 10,000 experiments that will involve individuals, community groups, schools, businesses, government departments or whatever – all working by appreciating what they already have and focusing their energies on reducing energy use and waste.

We want to replicate the models of renewable energy installations like the wind turbines in India. It might be biomass units in an African country or solar energy in Arabia. The world we have constructed, like all systems, relies on energy. We face social collapse without power; to avoid that collapse requires a transition to renewable energy as fossil fuels will run out. In some strange way we must thank our creator for climate change that is forcing this transition to happen now. If our fossil fuel use does ultimately converge on a global equal per capita amount with which the Earth can cope, then the energy from the sun is all that we will have to keep our complex societies going.

We want to help community groups form, develop and extend their influence to affect changes towards sustainable livelihoods. Our role is as support; we do not intend to lead, but rather follow where this grassroots approach take us. If it does influence policy and decision making, we will be there to help.

We want to forge links between communities to encourage learning about different cultures and values. As we build our more material connections to SCAD and similar organisations, we intend to exploit any opportunity for educational, artistic and spiritual exchange. Convergence of ideas about how we should live on this planet is a process of continual learning. One of the most exciting ways is through the performing arts, through drama and film making. We would like to develop ideas for plays and for the exchange of video diaries – joint productions across the world.

We want to use information technology to best effect. The Internet allows us to communicate directly and instantly. There are many opportunities to establish dialogue and to exchange large quantities of information, so using technology in the most effective way is another challenge. We will need to explore technical possibilities and move carefully towards opening sustained communication.

We want to research ways to create other convergent mechanisms. Carbon offsetting acts to transfer wealth from those who emit more to those who are below a fair per capita share. We would like to explore possibilities for balancing other 'allowances' extended to other resources. Is the idea of water offsetting one of these possibilities? As our global water resources come under strain, we may find this is a more pressing social concern than climate change. Should a company that depletes water to manufacture goods for export start to account for the full environmental and social costs?

The village of Chew Magna and the area in southern India are two systems that are divided geographically, socially, culturally, economically, and in many other ways – both tangible and intangible boundaries exist.

The two communities are part of a global system that is wedded to the traditional concepts of expansive growth. The gap in wealth between them is growing, the communities are diverging economically. However, due to advances in communications some aspects of culture are converging already – it is easy to imagine Manchester United supporters in Tirunelveli, and people in the UK appreciate Indian food and culture.

Chapter 3
Towards Convergence

"As the traveller who has once been from home is wiser than he who has never left his own doorstep, so a knowledge of one other culture should sharpen our ability to scrutinise more steadily, to appreciate more lovingly, our own." – Margaret Mead, *Coming of Age in Samoa*

There are thousands of initiatives to forge links between communities in different countries, from simple town twinning schemes to deep links created from years of personal contacts. Many of these initiatives adhere to a 'one world' principle, the concept that there is only one world of which we are all equal citizens. It is a compelling label, and many organisations have 'One World' in their name. There is OneWorld UK, part of the OneWorld International Foundation, which is a civil society, online network to help build a more just global society. There is One World Action that helps to transform lives, and to challenge the international policies that make and keep people poor. Linking peoples to fight injustice, to promote peace and equality through educational and cultural exchanges is an established tradition based on a simple approach that greater understanding will lead to solutions.

In contrast, The Converging World may seem a complex story: it contains quite a few components to address many problems. It is complex because it integrates a variety of solutions so that the sum of the parts is greater than the whole. This holistic and practical way of responding to climate change, resource depletion, poverty, injustice and inequality could hardly be simple.

What can we learn from other practical attempts to form links and to address issues of inequality and injustice? Many thousands of organisations and projects establish links between communities and address environmental and development issues. They are often combined with an agenda of conflict resolution and peace-making. This is a rich resource that has many lessons for what we would like to achieve in the context of Convergence.

A review of community-linking projects

The UK One World Linking Association (UKOWLA) is a membership organisation for more than 300 community groups in the UK. It was set up in 1984 to support, promote and encourage communities in the UK to develop partnership links with communities in the South.

The experiences of UKOWLA serve as a model for the processes of linking and an exploration of the issues that arise. They state clearly their beliefs:

- creative links between the peoples of the world through exchange, mutual practical support, through development education and advocacy are crucial for world peace

- these links enable people to see themselves and their responsibilities in the context of a diverse, complex and interdependent world

- mutual respect for cultural, social and religious diversity is a basic tenet of partnership, and that true and lasting friendship is a prerequisite for the promotion of understanding between the people internationally

- links to other parts of the world can and should be a powerful tool in creating greater social cohesion between diverse communities in Britain.

UKOWLA always stress that building partnerships between communities is a sensitive and slow process. Their experience is that you need proper regard to manage expectations and develop trust.

You should listen carefully, and try not to think about what you want out of the process but what you can give back. Do not presume to know what another community needs. Don't assume you know the answers to their problems; they might just as well know some answers to yours.

Realise that communications may be difficult. Silence doesn't mean they don't care, it may mean that the post is late or the wires are down. Don't assume similar systems and infrastructure are available; find out what does work. There is strength in numbers: involve as many people as you can, as the loss of one crucial person can break a link. Face-to-face meetings are crucial at some stage, despite the environmental costs. Remember you can't buy friendship with money or goods.

These and many other detailed practical lessons have been learnt over many years of supporting the development of links in many different countries. It is important to appreciate different and sensitive perspectives based on past experiences. For example, at one point UKOWLA say:

> "Expect your partner to be aware that they were once a European (perhaps British) colony and to have a different view of the partnership to yours based on the historical relationship."

One message that comes clearly from UKOWLA's experiences is the need for critical reflection on all sides, particularly on personal values and beliefs that are often emotional and controversial in nature.

They advocate a continuous process of exploring differences in knowledge, traditions, values and beliefs. Many issues are culturally embedded such as racism, violence against women, treatment of the elderly, and attitudes to sexuality. Often the simplest customs and traditions of one society may cause offence in another. Recognising acceptable and unacceptable practices, especially in religious contexts, is vital – getting things wrong without care can break relationships.

However, UKOWLA counsel against avoiding issues because they might cause difficulties. By not exposing differences you may reinforce power structures and prevent alternative voices from being heard. Despite all these pitfalls, it is always important to remain true to universal principles such as human rights. UKOWLA looks after the operational side of BUILD ((Building Understanding through International Links for Development), which is a coalition of 50 international agencies who work with politicians at government and UN level to bring community partnerships into the mainstream. BUILD has the goal to cultivate sustained community 'twinning' partnerships for learning between people across the world based on respect, trust and humility.

So what are the key principles for linking that we can gain from these years of experience from so many organisations, and what lessons do they hold for The Converging World?

Respecting and valuing differences – Convergence is not about creating one homogenous culture; it is the opposite. It is about allowing diversity to flourish in a secure society. However, it is important to raise awareness and advocate for universal concepts of human rights – civil, political, economic, social and cultural.

Equity and joint decision-making – Convergence is not a process that is imposed from one side. It is important that any framework or agreements for linking are made together with genuine dialogue and that the benefits are balanced and shared.

Make joint investments in the people and communities directly involved in the link. The Convergence process is often delicate and vulnerable. Appreciate what exists and build on it; particularly encourage women's groups and the marginal sections of the communities. Engage the young, the faith groups, and the existing local institutions and authorities.

Any process of Convergence should be conducted with a clear set of values. Many are obvious: open, honest, transparent motives, trust, dignity, mutual understanding, sacrifice, reciprocity and accountability. Successful links are dynamic rather than static, and they define many of their values, such as equality, from experience and not from some preformed theory.

Linking should not be a narrow end in itself; rather it should be a means of helping people to grow in awareness and understanding through real relationships and joint endeavour. Convergence, in the form of security, justice, prosperity and peace emerges over time.

Linking initiatives to reduce conflict

If there is anything further from a sense of a converging world it must be when two or more societies are in conflict. If we believe the forecasts of the effects of climate change then the future will hold many more conflicts, as resources such as food and water become scarce and people have no option but to migrate.

It is instructive to consider the peace movements for an understanding of the forces that prevent Convergence. Let us take as an example Peace Direct, an organisation established to support grassroots peacemakers, those involved directly in conflict areas. Its work consists of mediating, training mediators, and educating communities in alternatives to violence. The real essence of what it does is to be there with the message of non-violence, both on the spot and back here in the UK trying to influence people and government. It is always developing processes for mutual understanding.

One of the major initiatives of Peace Direct is called the Link Group Programme, which connects communities in the UK with groups in conflict areas. The objective, apart from moral support, is to bring the local peace work to the attention of the media and governments. By publishing details of conflicts and campaigning for change they hope to affect UK government policies. Processes of mutual learning and exploration of differences in cultures and expectations lead to much greater understanding of what drives conflict, and can lead to resolutions.

What can we learn from this programme?

Peace Direct stresses the importance of research to learn about the context and environment of the conflict area. It is important to understand who the key players are and their interests, the political structure and process, and the local, national and international policies that influence the conflict. The research should consider the history and the legacies that are still current, environmental and social factors such as diseases like HIV/AIDS, resource scarcities, refugees, communications; and should bear in mind what organisations are working in the area. When groups in the UK are linked to groups in conflict areas there is a focus and a purpose beyond mutual understanding. It is a difficult situation in which to promote the concept of a converging world, but placing this as an ideal may provide a new perspective to view conflict.

There are many other notable initiatives that are working in the spirit of Convergence:

The Ministry for Peace – this is a campaign for the creation of a new ministry within government and an independent civil society body working alongside it. "To work for peace is to work to transform violence. The fundamental aim of a Ministry for Peace is to reduce violence, both in the UK and internationally."

If we also consider the efforts of faith groups that have worked for decades at forging personal and community links, then we can appreciate an enduring and active desire to bring people together. The days of the Christian missionary attempting conversion may still leave a legacy in many countries, but the missionary spirit is now of mutual understanding and true coming together.

Other initiatives for Convergence

In building The Converging World we will learn from many organisations and movements that already exist. We intend to offer support and to blend in with processes that are clearly working to reduce carbon emissions and the use of resources in the demanding world, or are working for equity and justice in the processes of achieving Convergence. There are many examples, but we would like to pick out two.

WWF One Planet Living

WWF was founded in 1961 to tackle issues of habitat destruction and hunting that threatened Africa's wildlife. By the end of the 1970s, the focus of the work had broadened to include the wider implications of human activities on the environment. In 1980 WWF's World Conservation Strategy warned that humanity had no future unless nature and the world's natural resources were conserved. It also introduced the concept of sustainable development – living within the limits of the natural environment without compromising the needs of future generations – this has been central to their thinking ever since.

One Planet Living means what it says: we have only one planet to support nature and humanity. It's an initiative that aims to create a world in which people everywhere can lead happy, healthy lives within their fair share of the Earth's resources. The initiative is based on ten principles:

1. Zero Carbon – minimising CO_2 emissions from heating, cooling and powering our buildings

2. Zero Waste – minimising waste and flows of waste to landfill in a resource-efficient society

3. Sustainable Transport – reducing the need to travel and providing sustainable alternatives to private car use

4. Local and Sustainable Materials – optimisation of use of materials, in terms of their source and their performance

5. Local and Sustainable Food – maximisation of opportunities for use of local food supplies

6. Sustainable Water – minimisation of water consumption

7. Natural Habitats and Wildlife – protection of the natural environment and the habitats it offers to flora and fauna

8. Culture and Heritage – protection of the cultural heritage and the sense of local and regional identity

9. Equity and Fair Trade – promoting equity and fair trade within the community

10. Health and Happiness – promoting well-being and healthy lifestyles

WWF's aim is to build a worldwide network of One Planet Living Communities that demonstrate these ten principles in action and show that sustainable communities can work in very different locations. They aim to build communities in six countries across five continents, and to have started construction on at least two communities by 2010.

As we track the development of WWF from the early 1960s through to this new millennium, we notice the increasing emphasis on a complete reorientation of human lifestyles. Without this fundamental change there is no hope for conservation.

Fairtrade Foundation

The most widely recognised definition of fair trade was created by FINE, an informal association of the four main fair trade networks (Fairtrade Labelling Organisations International, International Fair Trade Association, Network of European Worldshops and European Fair Trade Association). It states:

> "Fair trade is a trading partnership, based on dialogue, transparency and respect, which seeks greater equity in international trade. It contributes to sustainable development by offering better trading conditions to, and securing the rights of, marginalised producers and workers – especially in the South. Fair trade organisations (backed by consumers) are engaged actively in supporting producers, awareness raising and in campaigning for changes in the rules and practice of conventional international trade."

Fair trade is about better prices, decent working conditions, local sustainability, and better terms of trade for farmers and workers in the developing world.

The fair trade organisations are increasingly successful examples of convergence in action. They address the injustices of conventional trade, which traditionally discriminates against the poorest producers, by getting a fair price, above the market level paid for their produce.

Fair trade is a voluntary model of trade that brings consumers and producers together. The difference in price paid between a fair return and a market-driven minimum is enough to stimulate virtuous circles – the natural forces for convergence. With some spare income, farmers send their children to school and can pay for better health care, leading to the next generation being more productive and better able to continue improving their livelihoods.

Labelling goods in the shops in the demanding world is a link, a direct connection, between the consumer and the original producer. It may be a fleeting connection, as the jar of coffee is bought, but taking responsibility for others in this tiny way changes your attitudes and behaviour to all the processes that support the lifestyles that we used to take for granted.

Of course fair trade, like carbon offsetting, has proved controversial. It has drawn criticism from both ends of the political spectrum. Some economists and right-wing think tanks see fair trade as a type of subsidy that impedes growth. Segments of the left criticise fair trade for not challenging the current trading system – again, perpetuating rather than replacing it.

This type of criticism is something that The Converging World will face; TCW is firmly in the camp of rapid evolutionary change – reformation. We don't have the luxury of time to tear down all the institutional structures and start again. Apart from the lack of time, the lessons of the last century are that you end up with something worse.

However, is it too late for reformation?

How robust are community relationships?

"More and more people, whose work I respect, are now out there saying it's almost certainly too late, in that we have crossed a number of critical thresholds on climate change and ecosystem degradation." – Jonathon Porritt, *Resurgence* Magazine

If we accept the rhetoric and logic about global warming and also peak oil, water scarcity, population growth, resource shortages and umpteen other miseries, then we need to prepare for a difficult future.

During a recent speech, United Nations Secretary General Ban ki-Moon stated that climate change poses as much danger to the world as war. This is, to many people, an understatement. Throughout 2006 a steady rise in the number of articles in the media about climate change turned into a flood. From the popular red tops to the slightly more mature broadsheets the potential problems of global warming were aimed at the general public. The alarms bells were rung, and everyone (especially organisations seeking funds) jumped on to the climate change bandwagon – except for those who are jumping off altogether.

The reaction of most people to an uncertain future is uncertainty – a wait and see attitude, as you can't believe everything you are told. Don't forget we are resilient, adaptable, and in true British fashion we will muddle through but only when our backs are to the wall. This is the apathetic response.

Some people deny there is a problem, saying global warming is a natural variation and a phenomenon that will reverse or at least will have as many beneficial affects as detrimental ones. This is the in-denial approach.

There are a few who welcome the apocalypse – the less said about them the better. This is the rapture approach.

Others recognise and believe in an apocalypse but they mean to survive it and emerge into a new world that has many fewer people in it. This is the survivalist approach.

At the far end of survivalism are those who are honing their woodcraft skills, living off the land and adopting an atavistic approach to preparations for the future. They foresee a return to mediaeval ways and even hunter gatherer societies – possibly still linked by the internet.

What is of concern to The Converging World is not this far end, not those who are really constructing Arks, but more the effect this survivalist attitude has on our perceptions for the future and how it shapes our intentions towards other cultures and people far away. We see signs of this as we build the modern ghettos, the gated communities designed to keep other people out and our thoughts inside.

Survivalism will lead to a closing of minds and a cutting of links. It will make the task of converging much harder as it directly reinforces the divergent processes of wealth inequality by hiding the realities from view.

Travel will become difficult as links between communities come under strain and mental frameworks contract, with people only focused on their immediate local problems. Already there are cries of hypocrisy about leaders like Al Gore who has done wonders for raising awareness of the issues, but is castigated for travelling the world by plane.

We have learnt, from the experiences of organisations like UKOWLA, that face-to-face meetings and cultural and educational exchanges are vital for newly formed and vulnerable community links. How do we maintain long-distance relationships when criticised for air travel?

Justifying why we want to invest in renewable energy in far-flung parts and setting up cultural links will become difficult. Already we have people saying 'Why not build wind turbines here, we'll need them when the oil and gas stop flowing.' Survivalism is short-term, linear thinking that is appropriate literally when your life is at stake. It is why one billion of the poorest people on the planet cause environmental damage. You will always cut down the last tree when not to cut it down means that you die.

Convergence means to resist strongly all ideas of survivalism. Our beliefs are in reinforcing the processes that draw us together as one people on one planet. We will continue to strengthen links and we will continue to travel to do so – in the most appropriate ways and making sacrifices elsewhere.

Let us also glory in a spiritual connection with the Earth when we travel to appreciate the vastness and the beauty, to experience the wonders like the colours of the coral reefs, the teeming richness of the rain forest, the silent landscape of the Arctic, or the skyscrapers against the setting sun.

Convergence as an ideal for linking

We have related the examples of the UK One World Linking Association, Peace Direct, WWF One Planet Living, and the fair trade movements. We are concerned about the strength of the connections they make. We have explored them in an attempt to learn as we construct our meaning for The Converging World. The question is: should we promote the concept of Convergence as an ideal for linking communities?

The existing frameworks that define reasons for linking communities are based on many different concepts. Often these frameworks are for conflict reduction – by encouraging mutual understanding, we reduce the potential for misunderstandings. Sometimes the frameworks are for social enrichment – cultural exchanges, awareness and education leads to a deeper and fuller appreciation of the world. Other frameworks are more explicitly for poverty reduction – again, education and the transfer of skills and technology are the motivators for developing links.

Is it too ambitious to propose that Convergence (note the big C) could act as a meta-framework, something that unifies these more focused frameworks?

The Converging World has grown into an idea that is holistic, which is why it is complex. It deals with the interconnection of the profligacy of the demanding world with the poverty of the developing. It embraces environmental pollution, energy for the future, linking communities, grassroots issues and government policy.

Can we develop a clear concept called Convergence that will make a difference? Can The Converging World embrace all that Convergence might mean? It may be a long time before we can attempt to answer these questions. However, asking them is a challenge to develop a programme of inquiry over the next few years.

One people – one planet – one chance.

Chapter 4

A very short history of Convergence

Let's step back in time and perspective. If we consider Convergence as a process to create one people on one planet, then can we identify it in the social and economic development of the world since the rise of modernity from the 16th century onwards? How does Convergence relate to our desire for sustainability, whatever that might mean?

Alex Inkeles (1998), in his book *One World Emerging?*, considers two aspects of convergence. One is the way that our institutional structures develop in similar ways despite very different political and cultural beginnings. The other aspect of convergence is in our popular attitudes and values. He suggests that these aspects are interrelated, and indeed that values and structures will co-evolve to similar points in every country.

A completely separate community that is self-contained and is self-sustaining is called an autarky. Some examples of these would be native Indian tribes that still exist, isolated in remote jungle areas, or maybe Eskimo tribes fifty years ago. But autarkies on a large scale have existed recently: for example the Soviet Union, Japan, Communist China and North Korea have all in their history closed their borders.

However, even on such large scales the interconnectedness of countries has intruded and made it impossible for this isolation to continue. Trade, communications and the exchange of ideas and cultural perspectives define this interconnection. At some point a degree of dependence arises: certain countries and societies start to rely on others. Trade with some other countries becomes vital; without it severe problems would occur, even social collapse. For example, the UK is totally dependent on a country like India for tea.

After a while, the economic processes of comparative advantage leads to an interdependence of regions and countries – a mutual dependence develops. Vital demands of all countries in some grouping are interlinked in a web of dependencies. Even the United States, which would still just be able to become an autarky by pulling up the drawbridge on the rest of the world, is increasing its dependence on Middle Eastern oil-producing nations.

Integration is the next step in the interconnectedness of nations. This occurs when vital functions and institutions are given up to a larger unit that comprises multiple nations. The European Union is now the body that sets laws and governs many economic and cultural aspects for 25 different countries. The US is an integration of the federal states.

Integration often leads to similarities in political forms, social organisation and cultural patterns, and we see the processes of an emergent global interrelatedness that we call convergence.

Not all forms of integration do exhibit convergence. Integration may well happen without convergence. For example, colonial dominance was a forced integration, a hegemony, which did not lead to convergence as divisions into a 'them and us' were maintained, sometime violently.

The history of modern industrial societies is one of increasing levels of convergence on five different dimensions:

Modes of production and patterns of using resources

This is the increasing use of the fruits of science and technology, driven by the impact they make on economic life. All societies have become dependent on the use of technology which is highly standardised. This is probably the major force for structural and economic convergence.

Institutional structures

A fairly standard way of organising and controlling our societies emerges often as a consequence of technology. In commerce we have similar forms of legal procedures, accountancy, insurance and taxation. In government the same types of ministries regulate and legislate in a similar bureaucratic (and labyrinthine) way.

Patterns of social relationships

The way that we develop class and social clusters and hierarchies is also a consequence of the divisions of labour required by technology and bureaucracy. Professional, artisan and unskilled classes emerge with elites at the top either through highly selective processes or as self-sustaining traditional ruling groups.

Family structures are also converging across nations: the so-called nuclear family that seemed a norm for societies is giving way to what is called the post-modern family, which describes the greater variability in family forms, including single-parent families and child-free couples.

Popular attitudes, values and behaviour

As people live similar lives governed by similar technologies, institutions, and patterns of social organisation, they start to adopt similar attitudes and norm of behaviours that fit within the stratification of the societies in which they live. So lawyers in one country develop similar values as lawyers everywhere. However, the values of lawyers differ greatly from those of petty criminals, so as long as we have an increasing differentiation of sectors in societies we won't all end up in one homogenous value set.

Systems of political and economic control

The processes of convergence may well, after many centuries, result in a great similarity of political and economic control. We may have had some 'jumping of the gun' in declaring an end of history and a triumph of liberal democracy. Yet there is belief in many circles that this is the desired outcome for all nations.

Inkeles identifies one aspect of political convergence that is truly common to all the states of the modern world and that is the growing power of the state to control the lives of its citizens:

> "[There is] a new tendency in the progressive penetration of systems of central control into more and more aspects of individual life. Whether this is a long-term trend towards which all humankind is converging, or whether the process is reversible in countries that attain a sufficiently high degree of development, I find very difficult to decide."

There is of course an enormous range of stages of development within and across nations, and convergence to some completely uniform world culture is highly unlikely given the resilience and adaptability of traditional cultural patterns and historical legacies. The vibrant and creative nature of humans means that new traditions form when societies are free from oppressive restraint. The bubbling forces of divergence in the form, interpretation and presentation of values will prevent bland uniformity, yet our convergent technologies will enable them to be shared instantly around the globe.

The five dimensions of convergence that we have considered so far are scientific and technological processes, social organisations, relationships, values and behaviours, and political and economic controls. What The

Converging World suggests is that another dimension, called global consciousness, is vital: in particular, an awareness of the predicament we are in and the need for collaboration on a scale never seen before, a collaboration that involves all countries and all people.

Convergence with global consciousness is a directed process, a deliberate way to alter the other five dimensions we have considered towards a common purpose of the survival of humankind.

Many countries are racing towards the modern industrial state and their economies are growing faster than at any time in the past. The world is linked together intimately to a degree never imaginable even two decades ago. All this economic growth is entirely dependent on fossil fuels. We can now understand the consequences of this dash for growth and how convergence towards this unsustainable ideal is logically doomed to fail. We only have a few decades to reorient our convergent processes towards a different ideal.

Globalisation

The forces of change, often lumped under the heading of 'globalisation', have produced rising living standards for some but a growing divergence of wealth within societies and among nations. The drive towards markets and democracy has led to increasing inequality and ethnic tensions.

However, it is difficult to disentangle convergence and divergence in the seemingly contradictory nature of globalisation.

There is a dichotomy in the phenomenon of globalisation that runs roughly along the lines of 'what are the benefits, and what are the disadvantages'. Classifying effects this way depends on your viewpoint. To the 'money men', the benefits of free trade, markets, instant communications and travel are obvious in driving economic growth and prosperity. To the landless poor, or the small farmers stuck with impossible debt, those 'benefits' seem remote and irrelevant.

This polarisation of views develops into the concepts of exploitative globalism, where the economic power of the rich countries, dominated by multinational corporations, expands voraciously across the world; versus benign globalism that sees the benefits of a convergence of economic development and institutional frameworks leading to common values and purposes.

This polarisation of globalism is matched by a polarisation of antiglobalism. On the one hand is an aggressive, destructive form of anti-globalism that resists the forces of exploitation, or seeks to destroy the systems that produce globalism. On the other is a constructive anti-globalism, a good globalism perspective that seeks to create a non-violent dialogue for a balanced global-local system that removes the exploitation, develops and protects diversity, and provides a pathway to equity and justice. This is exactly the philosophy of Schumacher.

It is clear that The Converging World sits in the paradoxical location of supporting benign globalism and constructive anti-globalism at the same time. Nobody said it was easy, but take heart from Felipe Fernandez-Armesto as he reflects:

> "The globalisation of culture is likely to be a self-defeating phenomenon. Whenever people get involved in big entities, they reach for the comforting familiarity of their local, regional or national roots. That is why superstates tend to break up after a while; and why old identities sometimes survive centuries of immersion in big empires. If the peoples of the world ever do come to think of themselves as sharing a single global civilisation, it will be a civilisation of a very heterogeneous kind, dappled with differences from place to place."

Sustainability

Sustainability is a driving concept in the work of The Converging World. We wish to encourage sustainable livelihoods both in communities like those under the wing of SCAD and in the 'Go Zero' communities of the demanding world. We use the term 'sustainable' loosely.

The underlying theories and notions of sustainability have been discussed and argued by many people. Our particular views are shaped by ideas such as Gaia Theory and the Limits to Growth work of the Club of Rome and The Natural Step. We live in an interconnected world that has finite resources, and the consequences of this are now familiar. These environmental and system dynamics perspectives demonstrate exactly what one world means for the reflexive dependencies of life and the underlying supply of goods from the physical biosphere. This gives us our saying that 99 percent planet living is the maximum we should ever reach – in contrast to the 300 percent we are unsustainably using today.

An agreed definition of sustainability is irrelevant. For the development of The Converging World we have more concern about the practice of making changes towards what we believe to be a more sustainable future. We follow Bryan Norton (2005) in his approach that puts an emphasis on co-operation and adaptation through social learning. This pragmatic and experimental way of working is most suitable when we deal with different communities, each experiencing unique situations in which they wish to make a difference towards some common purpose.

We hope that pragmatism linked to an appreciation of the convergence concept can direct groups. We believe in adaptive methods, slow steps, learning from doing and from the experiences of others. We wish to develop skills and practice in 'joined up' thinking. We also move beyond the idea that we are experts intervening in other people's lives, and firmly into methods of participation – a truly systems thinking approach.

The central tenet of sustainability is that we cannot have indefinite economic growth as defined by increasing consumption of finite resources. If we are to converge indefinitely to some ideal world, we need a vision that is logically possible. It must be towards a world that can support indefinitely the processes and continual changes in processes that make up life, and with us as an integral part.

Utopias, universal histories and the search for equality

"This moment of realisation, that it soon must be Utopia or Oblivion, coincides exactly with the discovery by man that for the first time in history Utopia is, at least, physically possible of human attainment."
– R. Buckminster Fuller

In the early 16th century Sir Thomas More described an imaginary island called Utopia where everyone lived free from oppression, poverty and war – a veritable place of harmony. Utopias are ideals constructed to explore political, scientific or economic values. Our concern with convergence is on the economic principles that we wish to transform.

Most economic utopian ideas were developed as a response to the harsh economic conditions of the 19th century. Early forms of capitalism were particularly oppressive, and exploited people in ways difficult to imagine today. A utopian socialist movement arose, with visions of an egalitarian distribution of goods, the abolition of money and fair share of produce for fair labour, with a strong sense of working for the common good leaving

plenty of leisure time for the arts and sciences. However, the socialist movement moved away from utopianism. Marx in particular became a harsh
critic of earlier socialisms he described as utopian.

Another type of utopian society was proposed by the libertarians.
These were capitalist ideas based on individual freedom and perfect markets. These individualistic and rather puritanical, work-ethic utopias saw
people as self-sufficient, requiring little dependence on collective action.

Many other economic and political grand narratives developed by
economists, sociologist and philosophers are non-fictional descriptions of
narrow utopian ideas.

All attempts to establish real utopias have failed. They were not only
premature and misconceived, but they were also exclusive. Small groups of
people withdrew from the rest of the world. Although supposedly independent and self-sufficient, they never were; they were constructed within,
and tolerated by, larger systems. The Amish communities and the kib-
butzim are probably the closest to any that have survived well.

Hidden away as much as possible, these communities either succeed or
fail depending on circumstances and, in particular, upon their beliefs
(Tobias, 2007). They act well as a retreat, but micro-utopias can only be a
temporary denial and comfort while that vast world-system, out there, continues its path to destruction.

Utopia must be, inherently, for all or none. No one could ever be truly
content in a utopia knowing that others outside suffered and deteriorated
while you, as a minority, prospered.

Should we describe convergence as a grand utopian vision?

Utopias are static places, like the biblical heaven on Earth; there is no
room for improvement or possibility of regressing. It's the end of the line,
and the journey is over. But convergence is dynamic, a continual searching
without a possibility of an end, because as fast as we can move together
there are forces, and a desire, to move apart. Our contention is that this balance is not sufficient at the moment and our attention should be on converging.

Back in 1980, in his book *The Third Wave*, Alvin Toffler introduced the
idea of Practopia. This was his vision of an emerging civilisation that is neither the best nor the worst of all possible worlds, but one that is practical. It
is a dynamic world, not a static perfect place. It is a positive, tolerant future
that makes allowance for individual differences, embraces racial, regional,
religious and cultural variety. It is a civilisation built in considerable measure around the home, potentially democratic and humane, in better bal-

ance with the biosphere, and no longer dependent of exploitative subsidies from the earth.

He imagines a world where every small effort, combined with millions of other efforts by people who struggle daily to imagine and build a more sustainable world, can multiply and bring us increasingly closer to a sustainable world.

This pragmatic approach is also reflected in R. Buckminster Fuller's ultimate design goal:

> "To make the world work in the shortest possible time through spontaneous cooperation without ecological offense or the disadvantage of anyone."

This is undoubtedly a model for The Converging World, but we are not suggesting that we will ever arrive at some perfect state, and it is quite feasible that on the journey we will often want to cast off the shackles of convergence for a while as we exhilarate in diversity and differentiation.

The concept of universalism is also relevant in our exploration of The Converging World. Wallerstein (2006) suggests that the intervention of the colonising countries of the West was justified by appeals to universalism – universal values of civilisation, development and progress. These were ideas that arose from the dialogue over the ethics of the Spanish conquest of the Americas. Such ideas were considered the natural law, and four assumptions were used to justify the interventions by the 'civilised' into the 'non-civilised' regions. These justifications were: the barbarity of others; violations of universal values; defence of innocents amongst cruel others; and the spreading of universal values.

However, Wallerstein argues that these concepts are not truly universal. Rather they were rationalisation for powerful counties to impose their will against the weak purportedly for their benefit and because it was historically inevitable by natural law. It was really only about power. However, those universal value arguments used then are still used now, often through the media and the activities of multinational companies. The assumption is that the Westernisation of the world is inevitable, and the end of history will arrive soon. This is becoming a minority view.

Wallerstein advocates a very different but true universalism. He demonstrates that there are universal values that demand critical appraisal of any justifications for intervention by the powerful against the weak. He wants us to seek answers to these questions:

"Can there be a universal universalism? Or rather, what would it take, in the twenty-first century, to arrive at a world where it is no longer the West that was giving and the Rest were receiving, one in which the West could wrap itself in the cloak of science and the Rest were relegated to peoples who had more 'artistic / emotional' temperaments? How can we possibly arrive at a world in which all would give and all would receive?"

Our job is to make sure that we follow this guidance and do not allow The Converging World to become the West giving to the Rest. There is a danger that convergence becomes another 'universal value', an excuse for intervention as part of some new game of power.

Our best defence against this danger is to embrace the ideas that we are all fellow citizens, there is no Us and Them, no distant other. These ideas are embraced by another 'ism', that of cosmopolitanism.

Cosmopolitanism is the concept that we all belong to a single moral community – we are world citizens. It overlaps with many of the ideas of universalism; it too promotes human dignity and rights that must be protected in international law. But it differs by explicitly recognising the differences among cultures across the world.

Cosmopolitanism is based on acknowledging the otherness of those who are culturally different, the otherness of the future, of nature and of other ways of reasoning. This acknowledgement does not extend necessarily to preserving these differences. Once again we are not arguing for the static only; there is a dynamic edge to otherness, and as convergence proceeds in one direction new forms of otherness open up.

A cosmopolitan view is one where the East becomes like the West and vice versa, where aspirations are universal and cultural differences are mild obstacles on the path to modernity. Both Eastern and Western ways of living have advantages, and the best ways will emerge as universal.

The cosmopolitan world view moves beyond the idea of nation states. The nascent global citizens and solidarity movements may form worldwide democratic institutions that allow appropriate local participation in decision-making – a cosmopolitan political order. The dynamic nature of our world is accelerating and the risks we face are getting greater in the impact they might have on our societies. We do face potential collapse if we make the wrong choices or fail to choose at all, but decision-making is not one side choosing for the other. As Ulrich Beck (2006) says:

"Global risks [like global warming] tear down national boundaries and jumble together the native with the foreign. The distant other is becoming the inclusive other – not through mobility but through risk. Everyday life is becoming cosmopolitan: Human beings must find the meaning of life in the exchange with others and no longer in the encounter with like. To the extent that risk is experienced as omnipresent, there are only three possible reactions: denial, apathy, or transformation. The first is largely inscribed in modern culture, the second resembles post-modern nihilism, and the third is the 'cosmopolitan moment' of world risk society."

This third, cosmopolitan way is the one adopted by The Converging World. It is bright green, optimistic and transforming; it acknowledges the other, it recognises and respects differences, yet seeks to reduce those differences that offend a universal sense of justice and equality.

Recent converging initiatives

"Half a million years ahead it may be a winterless world with trees and vegetation even in the polar circles. At present we have no certainty in such a forecast, but as knowledge increases it may be possible that our race will make its plans thousands of years ahead to meet the coming changes."
– H. G. Wells, *The Outline of History*, 1920

Ninety years after H. G. Wells's thoughts, we are nowhere near making plans for thousands of years ahead. What are the chances that we may make realistic plans for even decades ahead?

Many potential converging initiatives are showing mixed and disappointing results. The Earth Summit in 2002 is considered now to have failed. It did not produce many firm commitments, and those that were made had no chance of implementation.

Prior to the Earth Summit, in 2000 at the United Nations Millennium Summit, eight Millennium Development Goals were drawn up for the world's poor nations. These goals are to be achieved by 2015, and although they are certainly commendable they are vague and short on acceptable plans for implementation. The process is just seen by many as a neoliberal charter for the furtherance of privatisation. Few of the goals will be satisfied by 2015.

We have had G8 summits and Live8, we have noble grassroots movement like Jubilee 2000 debt relief and Make Poverty History – these efforts will continue. However, they have so far failed to galvanise the kind of coherent and total change in direction that the gravity of the challenges present. The media storm over climate change that built up in 2006 and continues now may allow some more far-reaching agreements, but not without significant resistance.

The Kyoto Protocol runs until 2012, when it needs to be renegotiated. There are hopeful signs that a son of Kyoto will emerge. However, will it be an expedient mix of ambitions and *realpolitik*, something that goes a little way in the right direction. Kyoto is a converging initiative that is achieving action, albeit not sufficiently to satisfy requirements for between 60-80 per cent emission reductions.

The Contraction & Convergence framework articulated by Aubrey Meyer provides for carbon trading between countries, those which are above and those which are below the limits set, on an equitable basis per head of population. This would allow developing countries to be compensated for suffering the consequences of the effects of carbon emissions by the most developed ones. It would also promote investment in carbon-free technology.

Sir Nicholas Stern told a large audience at the LSE in November 2006 that Contraction & Convergence was an argument that was "too difficult to get your head around". Certainly Stern can understand it, but why this reluctance?

Although Contraction & Convergence has had considerable endorsement from several governments (and many individuals), its full implementation would require massive changes to our lifestyle. As this is politically difficult, there have not yet been any serious attempts to implement it at an international level. The Converging World is an attempt to put into practice at a grassroots level the essence of Contraction & Convergence.

Solidarity movements

The definitions of community and solidarity are close: community is a body of people having the same interests; solidarity is a union of interests or purposes or sympathies among members of a group.

There are countless examples of protest movements across the world. Whether these are simple campaigns to redress injustice or complex and large-scale collectives that oppose oppressive regimes, they are linked philosophically into the concept of convergence. It goes without saying that The Converging World rejects all groups or individuals that do not adopt non-violent methods of resistance and protest.

Non-violent action, sometime called direct action, is often done through symbolic protest like demonstrations, but can be through non-cooperation and civil disobedience. Other strategies for social change may include information warfare, street art and theatre, boycotts, picketing, vigils, workplace occupations, and strike action.

These are frontline efforts for change, and can easily be seen as part of the process of convergence if they are to address issue such as inequality and discrimination. But equally they may be protests at the lack of effort by governments or others to preserve the environment or to put in place policies to tackle climate change.

The Converging World will demonstrate solidarity with people who struggle, non-violently, for convergent principles.

In our brief glimpse at an historical perspective for convergence we see it firmly rooted in the dialogues of globalisation and sustainability. It has a resonance of utopia, an ideal-seeking process where the end is perfection yet we accept that the end is not achievable – only the process is valid. We have looked at other ideas, universal and cosmopolitan concepts that reject the West versus the Rest and the continuation of imperialist desires, concepts that consider global citizenship and an end to the 'us versus the other'. A brief look at a few recent initiatives shows up a lack of progress and a lack of any real changes in worldviews. We have considered solidarity movements as vital aspects of convergence.

But what have we learnt? Perhaps one lesson is that an understanding of convergence requires critical thinking that comes through action and experimentation. Another lesson is that convergence embraces a wide range of ideas and philosophies, but can it provide a way to unify the many principles that have emerged to address the global crisis.

Chapter 5

Convergence as a unifying principle

"Never have so many systems vital to the earth's habitability been out of equilibrium simultaneously." – Lester Brown and Sandra Postel, *State of the World*

This is where we start to get theoretical – not that we see Convergence as a theory, for it is only a perspective. It is a viewpoint based on a collection of principles that we can apply as tools or as methods to evaluate actions. These principles that we are collecting have been developed over many years by many of the finest thinkers in ecology, sociology, politics and spirituality.

We are also adopting a survival worldview, which needs explaining. We are not survivalists in the selfish sense that we are only concerned with our survival and the continuation of our genes, family, or nation. Our survival worldview means that we expect the human race to continue for many millennia – at least until it has exhausted all possibilities for having fun and exploring creation. We expect this to apply to all peoples on the planet without exception. However, this is the century of crises and we face huge difficulties and events beyond our control. It may be an immense struggle to assure survival, let alone survival with justice and equity. The crises are certain to tax our moral values to the hilt. But, we, the people working to build The Converging World, expect humanity to survive if we have the courage to act.

By its existence as an organisation The Converging World is making a statement, but we know, because many people have told us, that the message is complex. It is an incomplete message and will take many years to unfold fully. At this stage there is little more than gut feelings and a compendium of the thoughts, aspirations and ideals expressed by countless people working in the fields of ecology, development, economics, human rights and justice.

The world in crisis – a systems approach

Let's start with a systems thinking approach to world problems, which explores the interconnectedness of human activity with nature. It should lead to an appreciation of the processes and feedbacks that shape the unfolding of our future, and examines what changes or interventions might produce converging effects rather than diverging ones.

A naïve, general proposition for Contraction & Convergence will act as an ideal. This is the basis of what we call The Converging World.

The world presents a scale of interconnectedness that is beyond our ability to comprehend. It is a world of feedback loops and non-linear processes that exhibit chaotic behaviour. It is a world of continuous change full of life and full of people.

In the Schumacher Institute we have found it useful to reduce all this complexity to just 12 'nodes' around a circle – see plate 8 in the colour section.

The bottom five nodes separate out the natural world: these are the physical, chemical and biological processes which, driven by energy and food, comprise the eternal struggle to reverse entropy – something we call life. In a sense, everything else that humanity constructs rest on these five nodes. Over the long period of human and social evolution we have created complicated and complex institutions out of the five basic building blocks. We build places to live and we build relationships, we exchange goods and ideas, we impose order and rules to look after our welfare, and ultimately we are able to perceive all that we have created and make some attempts to understand it. All this we have compressed into the other seven nodes: habitat, society, economy, trade, well being, governance and consciousness.

Although you may think that this is reductionism – splitting the world into 12 components – the way we work with this 'mandala' is to consider the interconnections. We have used it to look at the crises that we face; each of the 12 nodes contains a perspective on a different crisis. For example, in the hydrosphere node we have a crisis called 'the depletion of major aquifers beyond replenishment rate'; in the well-being node we have another crisis of 'human health problems caused by environmental pollution and toxicity'. We now inquire into the interaction of these two crises and potential ways to alleviate them. As we inquire we consider how the other 10 nodes affect our focus of attention.

When the principles and values of convergence are applied within the inquiry we have a direction, an ideal towards which we can attempt to move. In this case, what does it mean for our use of aquifers, what should we converge on, and what changes might shape this convergence? The ideal may be that we reach a point where aquifer depletion and replenishment are equal and we are able to allocate sufficient water supplies for use to cope entirely with human needs and the pollution that causes human health problems. Of course a full inquiry into an area like this is a colossal work, but it is not necessary to complete that before taking action. If what we do is governed by principles of convergence, our actions can be local, small and specific to what is happening now.

Each of the 12 nodes affords many opportunities to explore convergence and each of the 66 connections between nodes likewise is open to convergence analysis. Our simplified (maybe over-simplified) way of trying to get hold of the holistic nature of humans in the world is a first step to a more joined-up way of thinking about change.

The types of changes that we associate with convergence are those that bring things together and diminish the forces that are leading to inequality and divergence.

Processes that reinforce themselves are the expansive forces; they are essential in periods of growth, but they require mechanisms to rein them in as systems mature. Divergence is driven by positive (increasing) feedback: for example it becomes easier to make money once you have money. The larger a company becomes, the easier it is for them to reduce prices through economies of scale and that leads to a larger market share which leads to an even greater ability to reduce prices, and so on. We end up with a handful of giant multinational companies with revenues greater than many small countries combined.

Processes that diminish or control expansive growth are vital for maintaining stability and absorbing changes. Convergence is negative (diminishing) feedback: it is like the thermostatic controls that regulate room temperature. Without these, a boiler will continue to work until we swelter or the air conditioning keep going until we freeze.

The work of many people, like those behind the Limits to Growth, has clearly shown that we have too many reinforcing, divergent processes and too few convergent ones in our complex world system.

As Jeffery Sachs (2005) remarks:

"As the basic underlying forces that propelled the industrial revolution were replicated, multiple sites of industrialisation and economic growth took hold. Like a chain reaction, the more places that underwent change, the more they interacted with each other and thereby created yet more bases for yet more innovations, more economic growth and more technological activity."

Much as Schumacher called for appropriate scale, what we are calling for is an appropriate balance of processes: enough reinforcement to provide the dynamic leading edge of technological innovation with enough constraints and diminishing processes to maintain a stable whole.

The study of convergence will occupy us for many years to come. It involves an extremely wide range of subjects and necessarily requires a sense of convergence in the academic world.

Deep Ecology as a set of principles for convergence with nature

"Humanity does not consist in the fact of survival and eventual death, but reveals itself in how we survive and die. Unlike flowers, leaves and grass, we experience death first and survival afterwards, and for us survival represents a choice. What is man? An animal that can choose to survive – for a while." – Walter Kaufmann

Proponents of Deep Ecology believe that the world does not exist as a resource to be freely exploited by humans. Following Arne Naess, a deep appreciation of nature, and the wisdom we can derive from that, should form the basis for an ecological ethics. How does this relate to our development of the concept of convergence?

Deep Ecology offers an eight-point platform for how we should view our place in nature:

1. All life has value in itself, independent of its usefulness to humans.

2. Richness and diversity contribute to life's well-being and have value in themselves.

3. Humans have no right to reduce this richness and diversity except to satisfy vital needs in a responsible way.

4. The impact of humans in the world is excessive and rapidly getting worse.

5. Human lifestyles and population are key elements of this impact.

6. The diversity of life, including cultures, can flourish only with reduced human impact.

7. Basic ideological, political, economic and technological structures must therefore change.

8. Those who accept the foregoing points have an obligation to participate in implementing the necessary changes and to do so peacefully and democratically.

If you accept the eight points of deep ecology, then you may change your perception of our place in nature. You will agree that massive human economic activity has pushed the biosphere far from its 'natural' state through reduction of biodiversity, climate change, and other influences. As a consequence, civilisation is causing mass extinction and is facing its own.

The deep ecology philosophy clearly supports the ideas within convergence, especially with its contention that the ecosystem can absorb only limited change by human activity. However, the emphasis on humankind as just one species on the planet with no exceptional status, and an implied anti-technology bias makes it difficult to square well with the emphasis that The Converging World places on social justice and a drive for appropriate uses of industrial processes.

This is a substantial area of philosophy that does form much of what is needed to support convergence.

Human rights as a set of principles for converging behaviour

"All of life is interrelated . . . whatever affects one directly, affects all indirectly."
– Martin Luther King Jnr

Implicit in utopian dreams is this sense of equality: all men are equal. However, our present economic inequality is more obvious than our political inequality. It is appalling that so many people are starving while others have more to eat than is good for them and more money than they know how to spend; but it is a far cry from this feeling to the demand that all men ought to be made economically equal. We are used to the call for political equality, yet extending that call to social, cultural and economic spheres has laboured badly.

Perhaps the Deep Ecology influence is a good combination to make with the human rights movements. Together they provide a more coherent platform to build the principles of convergence.

Human rights advocates may need to think more about environmental problems, like climate change, as rights issues. Deep Ecology may overemphasise the Earth, but part of the reason for the lack of progress in human rights may stem from the emphasis on 'humans' without sufficient concern for the environmental stewardship that underpins so many social and economic rights. The possibility of embracing environmental concerns more explicitly is clearly within the human rights framework.

In her Barbara Ward Lecture in 2006, Mary Robinson said that climate change raises fundamental issues of human rights:

> "Article One states that all human beings are born free and equal in dignity and rights. However it is poor communities who are suffering most from the effects of climate change, and it is rich countries that are contributing most to the problem."

The human rights approach, emphasising the equality of all people, challenges the global power imbalances that allow the main producers of carbon emissions to continue unchecked.

Human rights may provide legal as well as moral grounds for empowering the poor to seek redress. For example, in 2005 Inuit communities in Alaska and Canada filed a petition with the Inter-American Commission on Human Rights that the impacts of climate change, caused by the US, violate their human rights. The Inuit say their livelihoods, their spiritual life and their cultural identity are threatened by the US government's failure to curb greenhouse gas emissions.

Sheila Watt-Cloutier, a Canadian Inuit activist, said: "We offer our testimony as a warning to humanity that, while global warming has hit Arctic peoples first, changes are coming for everyone."

Let us focus on two of the articles in the Universal Declaration:

> Article 25: "Everyone has the right to a standard of living adequate for the health and well-being of himself and of his family, including food, clothing, housing and medical care and necessary social services, and the right to security in the event of unemployment, sickness, disability, widowhood, old age or other lack of livelihood in circumstances beyond his control."

Article 27: "Everyone has the right freely to participate in the cultural life of the community, to enjoy the arts and to share in scientific advancement and its benefits."

These two rights alone are solid justification for pursuing our concept of Convergence. Our continued existence on this planet demands that we establish a fairer way of sharing out the burdens and benefits of life, and that we hold as central values the rights of both today's poor and tomorrow's children. If we accept this, then it implies contraction in the demanding world and simultaneous expansion in the developing sectors.

Earth Charter

The Earth Charter is a declaration of fundamental principles for building a just, sustainable, and peaceful global society for the 21st century. It was created by a global consultation process covering thousands of people, and it has been endorsed by thousands of organisations representing millions of individuals. It inspires a sense of global interdependence and shared responsibility for the well-being of humanity and the whole living world. It is a unique blend of deep ecological principles with human rights.

As a declaration of where we would like to be it is unassailable. In its statement of the way forward there is a desire to gain endorsement by the UN, and to see the charter implemented in legally binding instruments on environment and development.

Alan AtKisson, the Earth Charter Initiative's executive director, is keen to describe the extremely broad and diverse movement that has grown up around the Earth Charter, with its differentiated qualities in countries all around the world.

The strategic work of Earth Charter International can be summarised in the closing words of the preamble:

"Therefore, together in hope we affirm the following interdependent principles for a sustainable way of life as a common standard by which the conduct of all individuals, organisations, businesses, governments, and transnational institutions is to be guided and assessed."

We have discovered strong influences for the concept of Convergence in deep ecology and human rights, so the Earth Charter combines and expands these to provide us with the values and principles we need to establish a framework for action. It is far from prescriptive, and can be dis-

missed as wishful thinking on a grand scale; however, without a clear statement of the ideal we have no guiding light and no mechanism for measur-
ing our progress.

The Converging World is fortunate to be closely associated with The
Earth Charter and will endeavour to realise the charter's goals.

For a simple example, take subsection 10b which says we need to
"enhance the intellectual, financial, technical and social resources of developing nations." TCW, like so many organisations, aims to achieve this by
funding the conditions for sustainable livelihoods with agencies like SCAD
and by forming direct community links. These actions, multiplied across
the world work directly towards our concept of Convergence.

Schumacher, and other principles and values for The Converging World

How does The Converging World fit with Schumacher's philosophy? To
attempt an answer, let's consider the following six key concepts that we
have distilled from Schumacher's work.

1. *Appropriateness* – human endeavours at the right scale for human ful-
   filment.

2. *Personal and local actions* – that multiply into the critical mass needed
   to create whole system changes.

3. *Integration of the inner and the outer* – finding ways to unite differences
   to create one world, global citizenship, and positive globalisation.

4. *Non-violence* – adopting gentle, organic mechanisms. This is not only
   directed at how we live together but also how we live with nature.

5. *Spirituality* – placing our work within creation. This is the understanding that systems are not 'out there' but that we are part and that we
   matter.

6. *Simplicity* – seeking to demystify and find uncomplicated explanations
   and solutions so that there are no barriers to exclude anyone from participating in creating a better world.

It is reasonably clear that numbers 2, 3, 4 and 5 are well represented in The
Converging World. Number 6 is an ambition, and we are working hard on

making our story less complex – we will make sure that all our work is as inclusive as possible, and that means appropriate language.

The issue of scale (number 1) is interesting, as we do indeed agree that our community projects should be the appropriate size and shape to meet community needs. The whole TCW project, however, may not be small scale. It is important to present TCW as a large-scale development as we need the credibility and profile that this would give in order to attract donors willing to fund expensive turbines to achieve the volume of carbon savings that we want. We may say that in this case appropriate may mean quite big.

The Natural Step

The Natural Step Framework's definition of sustainability includes four system conditions or scientific principles that lead to a sustainable society. These conditions are that in a sustainable society, nature is not subject to systematically increasing:

• concentrations of substances extracted from the Earth's crust;

• concentrations of substances produced by society;

• degradation by physical means

and, in that society. . .

• people are not subject to conditions that systematically undermine their capacity to meet their needs.

These conditions are fundamental to the way that we wish to converge. Another saying that we use to describe TCW is:

> "Connecting communities through action to maximise our quality of life while achieving equal and minimal impact on the Earth's resources."

The only way to achieve this is through principles such as these four conditions. Karl-Henrik Robèrt, the founder of The Natural Step Framework, describes how it would create change:

> "I don't believe that the solutions in society will come from the left or the right or the north or the south. They will come from islands within those organisations; islands of people with integrity who want to do something. . . ."

"This is what a network should do – identify the people who would like to do something good. And they are everywhere. This is how the change will appear – you won't notice the difference. It won't be anyone winning over anyone. It will just spread. One day you don't need any more signs saying 'Don't spit on the floor,' or 'Don't put substances in the lake which can't be processed.' It will be so natural. It will be something that intelligent people do, and nobody will say that it was due to The Natural Step. It will just appear." – Karl-Henrik Robèrt, 1991

This is the process we are following for convergence. Our islands start with agencies like SCAD and community groups in the west of England.

Spiritual convergence – a challenge to faiths

"Someday, after we have mastered the wind, the waves, the tide and gravity, we shall harness for God, the energies of love. Then, for the second time in the history of the world, man will have discovered fire."– Teilhard de Chardin

Although the quotation from Teilhard has a resonance of Francis Bacon's conquest of nature, it is more comforting to think of a time when we have blended into nature with our use of renewable energy – it may be fitting for a country like Britain that he didn't mention mastering the sun. But to harness the energies of love in the pursuit of convergence also has a sense of blending our actions into a faith in a creator.

Calls for inter-faith dialogue on environmental and social issues are going out across the world. The importance, to our study of convergence and the development of The Converging World as an organisation, is the potential that faith groups have to initiate changes at the personal and local community level. Every physically defined community in every part of the world has a worship group of some kind or another.

Some of the questions that require long periods of inquiry are:

- How do different faiths understand the interdependence of all life?

- How do we enshrine this understanding in communities? What is the role of faith in activating and motivating communities to make changes?

- How do people of different faiths comprehend their relationship with the Earth and its resources? How can faith communities contribute to sustainability?

- How can we integrate faith efforts for peace and conflict resolution with efforts to create sustainable livelihoods?

As an example of what is happening, consider the One Earth series of discussions held in 2006 that invited people of different faiths and none to explore and reflect on issues of global concern.

One Earth offered an opportunity to engage people from a diversity of backgrounds, beliefs and knowledge by bringing different faith perspectives to bear on current issues and exploring how people of faith can work together to face the challenges of a globalised society. Work of this kind provides a complete summary of The Converging World concept:

> "In the midst of a magnificent diversity of cultures and life forms we are one human family and one earth community with a common diversity."

> – Preamble to the Earth Charter

The call for Global Governance

"Everything in the world may be endured except continued prosperity." – Johann Wolfgang von Goethe (1749-1832)

Within Convergence we have embraced cosmopolitanism, the idea of citizens of one world with a common, human purpose. This takes us in the direction of global governance and the need for supranational institutions.

Our world is full of international trade, multinational corporations and astonishing communications technology such as mobile phones, the internet, and satellites. At the same time as these advances in economies and technology we have seen the growth of global institutions such as the United Nations, the World Trade Organisation, the IMF and World Bank and the International Criminal Court. We can add to this the development of regional bodies such as the EU, the rise of transnational corporations and integration of markets often termed economic globalisation, and the emergence of global NGOs and transnational social movements, such as the World Social Forum.

A cosmopolitan identity emerges, and we reach out for the concept of global citizenship to supplement our national identity. This globalising of citizenship is a consequence of the convergence of governance. As the institutions of the modern industrial state emerge, they increasingly conform to a common structure and function. They also increasingly require cross-border co-operation as trade and the free flow of people rapidly grows and states become interdependent.

However, the prospects for a world government still seem some way off. It is something that can arouse deep suspicion, especially to those who fear a preponderance of American influence on the global stage.

This aspect of convergence – moving towards a world government – is far from the ambitions of TCW, but we are closely involved with an organisation that is concerned with world institutions and functions. The World Future Council aims to create a long-term voice to speak up for future generations and to build a global forum based on shared ethical values. The World Future Council wishes to create a permanent international forum that will foster long-term thinking for today's world. TCW fully supports the Council.

Chapter 6

A critique of
The Converging World

"Criticism is something we can avoid easily by saying nothing, doing nothing, and being nothing." – Aristotle

Is The Converging World an appropriate response for a survival worldview?

This final section is a critical examination of the proposition for The Converging World and the concept of Convergence.

There are some rather practical matters on which we could get criticised, which raise questions such as: 'Are there any issues over the land on which the turbines are placed?' 'Were any farmers displaced or livelihoods affected?' 'What are the credentials of the company operating the wind farm?' 'Are we happy that no obvious exploitation has occurred?' 'How do we know what will happen to the turbines in the next 20 years?' 'Are we raising expectations with SCAD?'

These direct, practical issues are important, but they are all operational. Our response is that we will do our best. So far we believe that we are aware of all these risks and dangers, we have tried to think it through.

More difficult questions arise when we pull back to get the work into a wider perspective. A range of questions emerge like:

'Are we trying to be too clever?' 'Should we construct a charity as complex as TCW?'

'Is it serious? Is it secure? Will relationships endure if times get difficult?'

'What is the morality of Convergence, and what are the consequences of pursuing methods to encourage or even impose this way of thinking?'

Maybe to start this critique we should embrace the uncertainty and ambiguity that are inescapable in a postmodern, globalising world. This allows us to develop a humility, especially about what we know, which will counter the risks of moral and cultural imperialism often lurking beneath the surface of this universalistic, cosmopolitan and pragmatic project.

One main argument against this project is that convergence will exacerbate the negative aspects of globalisation. It will lead to a loss of cultural diversity, a bland homogenisation so that everywhere becomes everywhere. A cultural loss of identity and the creation of ersatz cultures is not part of the motivating spirit of this challenge; however it may be an unintended consequence unless we are alert to it happening.

Other classic economic refutations offer free trade and free market concepts as the answer to issues of poverty and underdevelopment. If lesser developed areas of the world are allowed the freedom to grow and trade then rising levels of income will introduce different behaviour, in particular a greater sense of protecting the environment. Why not just promote this approach?

Markets and the environment – carbon offsetting

The concept of The Converging World is firmly based in Contraction & Convergence and carbon trading. However, the arguments against carbon offsetting are getting louder, and they focus on (a) this is guilt money, (b) it is applied badly and causes detrimental effects on ecosystems and cultures, and (c) we are encouraging the development of a consumer society (the electricity encourages a 'Western' lifestyle), which involves economic growth and more energy use and carbon emissions.

We have addressed these particular and extensive criticisms of carbon offsetting in an earlier section, but the general issue remains that carbon trading is another example of commodifying nature and common goods; it is turning values and intangibles into products that can be bought and sold and traded with all the panoply of futures and hedging and mechanisms to make money for market traders.

Sharon Beder (2007) makes a clear proposition that market-based environmental policies are unable to properly protect the environment. She says:

> "Human rights are meant to be inalienable, which means they cannot be taken away, sold or given away. Yet economics-based environmental policies do just that. Access to environmental resources . . . and to a healthy environment becomes just more figures in the calculus of economics-based decisions."

Are we pursuing the wrong path by using market mechanisms to generate and multiply funds? What other options do we have?

A further serious criticism arises by stepping further back and realising that we may get locked into market-based solutions and even if everyone were to agree they were not the right answer, we would be unable to go back.

More criticisms flow: there are limits to what can be achieved by individual actions alone, so it is not convincing that this project is an appropriate response for a survival strategy; the existing economic systems so favour mass production and distribution on a global scale that it is difficult to buck the system and find other more just and life-sustaining ways of living and working together within the global family.

There are some good success stories – especially at the community level – where people have found ways of short-circuiting the global economy: community-supported agriculture, farmers' markets, and community currencies. But these are exceptions and somewhat marginal. This, of course, is why we struggle to make ethical consumption – organic, fair trade, etc. – truly socially inclusive. Many poor people simply cannot (or believe they cannot) afford to take part.

Expectations and language

We become open to criticism as soon as we begin to communicate our ambitions and plans with people who have different worldviews.

As an example, TCW is a charity and is not flush with money. Everything has to be accounted for scrupulously, and it lives with the expectation that a charity operates with the minimum resources possible so that it can funnel the real money to the greatest benefit. Elsewhere in the world people coming from the West are seen as always having money, so if they decide not to go ahead with a scheme then it means that they don't want to do it, not that it is difficult or too soon or doesn't quite fit with other ideas. We can raise expectations without realising it just by our cultural norms not matching.

We have talked about linking communities in the ratio of one to five by population to reflect the one billion rich people in the North/West/Demanding World and the five billion in the poor South/East/Developing World. Doesn't this reinforce a division, when we are looking to converge? Words are important when we describe the processes involved with The Converging World.

Linking

There are many tremendous twinning type relationships already. What can The Converging World add that is truly new?

Are we in danger of being patronising – the rich West with all the answers again?

We need to know and understand what the people from SCAD might want from the people of Chew Magna, and vice versa. What role would each community play in this partnership? What kind of exchange of ideas and initiatives would generate win-win outcomes for both communities involved? What will it take to make it happen and if so, how long will it take and what do we need?

Having alluded to participation and democracy, we will need to understand how SCAD works and present that to potential contributors so that they can understand how the communities will share in the processes. We need to demonstrate how decisions will be made and who makes them – a call for transparency and accountability.

Framework of ideas – what constitutes our motivation?

When we take time to reflect we ask the question: why are we doing this?

It is likely that the motivations for involvement in The Converging World are complicated and personal, and getting involved may not have been that deliberate. This is a process that has emerged from other activities, from making connections and exchanging ideas.

Some possible answers to the question 'Why are we doing this?' might be a sense of guilt, a desire to do good (or to be seen doing good), fear for our survival and our children's, or dissatisfaction and a lack in our lives – it may be all of these and more.

Is there a sense of power and control that needs revealing? This project is coming from a group of people who are well aware of the issues of climate change and global warming and are dedicated to pursuing solutions to these problems. Within this pursuit there are structures and relationships of influence – these may provide individual motivation.

The UK relinquished colonial control in India 60 years ago, and we went our merry ways. We now realise the mess that our economic growth is producing, and we are particularly afraid that if countries like India want what we have, then the problems become acute. Suddenly, we want to change things by entering a partnership and offering the prospect of a convergent world. Why should anyone trust us?

Continuing exploration of these questions is important as we try to understand what we are doing and as we try to understand alternative worldviews held by others when we enter into discussions across the gap of cultures.

After asking why we are doing what we are doing, we need to ask the question: ought we to do it?

When a group met in Bristol in July 2005 for a first exploration of the concept of The Converging World, it was roughly a mix of rich, older, white men and keen, young people full of energy and conviction. Conversation ranged around ways to create relationships between Chew Magna and SCAD in order to show how the principles of Contraction & Convergence could be implemented at a local scale. Only one person at that meeting had been to Tamil Nadu, but there were two residents of Chew Magna. The group wasn't really 'of' these places; its outside position encouraged a way of talking that was imposing ideas onto abstract concepts. It was a 'doing it to them' way of thinking.

Do we have a right to attempt this experiment? We may cause damage by unintended consequences, we may raise expectations that we cannot meet. Do we have the attitude that we are the experts and we know what changes are necessary – do we consider that our ideas are right for these communities because they are our ideas? Have we let our imaginations run too far ahead of reality?

The 'fallacy of misplaced concreteness'

The 'fallacy of misplaced concreteness' was originally stated by the philosopher Alfred North Whitehead. It involves thinking something is a concrete reality when in fact it is merely a belief, an opinion or a concept about the way you want things to be. It is akin to what we might commonly call building castles in the air.

This treatment of an abstract set of principles as though it were a concrete entity is a serious danger for new projects. You meet someone, you explain what you hope will arise one day but you slip into talking as if it were already constructed. They then go away, firmly believing and often misinterpreting as well, and later you find a newspaper report that clearly states that you have achieved remarkable things.

The fallacy of misplaced concreteness occurs when we mistakenly think that our visions and theories of the world are the world. But these imaginations are nothing more than tools to help us explore. They are not ends in themselves; they are means to an end, which is our understanding. This is not a simple process but one that evolves as our work develops and circumstances change.

Those who postulate a utopia usually fall into the fallacy of misplaced concreteness. Utopians develop a complete plan for what they believe is the ideal society, and then they work to make it a reality. This is harmless when done on a smaller scale. Such communities often don't last very long. However, any free society should have room for such experiments. It's at the heart of what we mean by freedom of thought and action. The danger comes when, on a larger scale, we brand those who do not agree with our aims as somehow standing in the way of progress.

The Converging World may be utopian to some, but we should resist this thinking. We don't know, and cannot know, where this project will end up. The ideals we may have should illuminate the path we take; but contraction and convergence are ideas not things.

Beyond carbon

The Converging World and our concept of convergence grew out of the efforts to tackle global warming. It is heavily influenced by a single element called carbon that is the basis of all life, but, in a simple molecule combined with that oxygen, another life-defining element, it causes a frightening problem.

Carbon has many personalities. It is a living substance when seen as a tree formed from the atmosphere; it is dead when dug out of the ground as coal; it is wealth when it sparkles as a diamond; and it is the invisible blanket that is warming us beyond our comfort zone.

Enthusiasm for carbon markets has sparked ideas for other schemes such as markets in 'wetlands banking' and 'endangered species credit-trading'. Many people refer to the trading of rights to emit sulphur and ozone-depleting chemicals as successful processes to control environmental pollution. There are schemes to buy rainforest and prevent logging. You can buy real and virtual goats, trees, water buffalo, wells, and children's education. The well-established Fairtrade movement moves money from consumers to producers better than any current market mechanism.

John Vidal (2006) extends the concept of personal emissions trading. He foresees the potential for resource economies developing in many other areas, especially water. This is going beyond carbon offsetting to deal with the rest of our ecological footprint:

> "A global hydro economy would demand that industry and farming would use less water, and would encourage different crops to be grown and fewer animals to be bred for food. . . . Tomorrow it will be much easier, cheaper and more convenient to switch to a green lifestyle."

By letting our imaginations run away we have suggested that after carbon we might consider ideas for voluntary water offsetting, for tradable fishing rights, salinity trading and offsetting. The most taboo idea would be to off-set children – if you want a third child then you pay money that goes to a population control agency. However wild these ideas become, they are all driven by the concept of a convergence of our economic lifestyles and our ecological footprint – reaching for that ultimate point where we are all equal but different in the impact we make, and the total impact is acceptable to nature.

We realise that all these attempts and crazy ideas may be turning the environment and rights into products for sale, which then forces us to rely on markets rather than democratic institutions for 'solutions'. There is great concern that this is not the correct approach. However, they are attempts to implement the spirit of economic convergence, which is to reduce consumption where it is over the appropriate level and transfer money to where consumption is below the appropriate level. The flow of money enables action.

Conclusions

"Contraction & Convergence is a mechanism by which money would flow from rich to poor nations as of right, not as aid. It would introduce a mechanism by which the world economy would move in the direction of greater equality. It is based on a simple principle of justice that any schoolchild can understand, and therefore should not be difficult for negotiators to pervert."
– James Bruges, author of *The Little Earth Book*

What we have explored in The Converging World is all beginnings. Every which way that we have looked at convergence is only skimming the surface of deeper ideas. We have set an agenda for discussion and action, and for many years of future experiment, action research and critical thinking.

For now, the conclusion of this Briefing is a challenge. Can you help us to make The Converging World a substantial force for change? There is no inherent prescription in what we are doing, nor will there ever be a prescribed path; we have presented the concept of Convergence as a pragmatic process towards a cosmopolitan ideal. If accepted as a challenge, we suggest that local action through linking communities is a vital method to achieve changes, not just because 'small wins' have intrinsic value but because personal actions and learning can snowball into larger movements that can make major differences.

There is nothing startlingly new in the idea of local action. What we present is an elegant way to combine ideas to get much more out than we put in. We have a cycle going: persuade > use > persuade > use.

1. Persuade the wealthy in the demanding world to fund renewable energy in the developing world. This saves carbon emissions.

2. Use funds from selling the energy to drive sustainable development in the developing world. This raises living standards.

3. Persuade everyone in the demanding world that they should contribute money to balance the emissions they cannot reduce. This allocates carbon saved.

4. Use the donated funds for contraction in the demanding world, through efficiency, renewable technologies and social change. This reduces energy demand and carbon emissions.

There are dangers and disappointments ahead. We need partners like SCAD to continue working with their communities to struggle for better livelihoods, to shake off oppression and take charge of their own lives. We must campaign for a change to the underlying philosophy of economic growth: contraction must lead to overall reduction in fossil energy usage – not freeing up supply for someone else to use. We must not give in to despondency or get overwhelmed by survivalism and start shutting the gates.

It is amazing what comes from a few people sitting around a table having a coffee-house conversation – the ripples stretch to the other side of the world, and lives change.

We have barely scratched the surface of what convergence means, but we have a vehicle to explore the territory. The Converging World will boldly go forward, led by the heart. The Schumacher Institute will follow, but lead by the head – trying to make sense. There is a huge task to integrate thinking – and to integrate action – behind this simple concept of convergence. There is an even bigger task to accomplish in building The Converging World.

If you feel persuaded that we might be on the right track then join us by becoming a member of The Converging World, we can use you.

www.theconvergingworld.org

Resources

References and further reading

Appiah, Kwame Anthony, 2006. *Cosmopolitanism: Ethics in a world of strangers*, Allen Lane.

Beck, Ulrich, 2006. 'Living In The World Risk Society', A Hobhouse Memorial Public Lecture given on Wednesday 15th February 2006.

Beder, Sharon, 2007. *Environmental Principles and Policies*, Earthscan.

Chomsky, Noam, 2002. *Understanding Power*, p388, Vintage.

'Carbon Trading: a critical conversation on climate change, privatisation and power'. September 2006, www.thecornerhouse.org.uk.

Dasgupta, Samir & Kiely Ray, 2006. *Globalisation and After*, Sage Publications.

Fernandez-Armesto, Felipe, 2000. *Civilisations*, Macmillan.

Fals Borda, Orlando, 2006. 'The North-South Convergence'. Action Research Vol 4, Issue 3, Sept 2006. Sage.

Fuller, Buckminster, 1972. *Utopia or Oblivion: the prospects for humanity*, Viking

Inkeles, Alex, 1998. *One World Emerging?*, Westview Press.

Jones, Tobias 2007. *Utopian Dreams*, Faber & Faber.

Keiner, Marco (ed.), 2006. *The Future of Sustainability*, Springer.

Kingsnorth, Paul, 2003. *One No, Many Yeses*, The Free Press.

Koch, Richard & Smith, Chris, 2006. *Suicide of the West*, Continuum.

MacGillivray, 2006. *A Brief History of Globalisation*, Robinson.

Ma'anit, Adam 2006. 'If you go down to the woods today . . .' *New Internationalist*, July 2006 (CO2NNED – carbon offsets stripped bare).

Meyer, Aubrey, 2000. *Contraction & Convergence: The Global Solution to Climate Change*. Schumacher Briefing No. 5.

Miliband, David, 2007, announcing public consultation on carbon offsetting.

Norton, Bryan, 2005. *Sustainability*, The University of Chicago Press.

Porritt, Jonathon, 2005. *Capitalism: as if the World Matters*, Earthscan.

Robinson, Mary, 2006. 'Climate Change and Justice'. Barbara Ward Lecture, Chatham House, London.

Sachs, Jeffery, 2005. *The End of Poverty*, Penguin Books.

Sacks, Jonathan, 2003. *The Dignity of Difference*, Continuum.

Sardar, Ziauddin, 1998. *Postmodernism and the other*, Pluto Press.

Sen, Amartya, www.removingunfreedoms.org.

Singer, Peter, 2002. *One World: the ethics of globalisation*. Yale University Press.

Toffler, Alvin 1980. *The Third Wave*, William Morrow & Company.

Toulmin, Stephen, 1990, *Cosmopolis*, The University of Chicago Press.

Vidal, John 2006, 'Green living guide', *The Guardian*.

Wallerstein, Immanuel, 2006. *European Universalism*, The New Press.

Websites

Carbon Trade Watch www.carbontradewatch.org

Commission for Equality and Human Rights www.cehr.org.uk

Debt for Nature www.worldwildlife.org/conservationfinance/swaps.cfm

Earth Charter www.earthcharter.org

Global Commons Institute www.gci.org.uk

Go Zero www.gozero.org.uk

International Futures Forum www.internationalfuturesforum.co.uk

Ministry for Peace www.ministryforpeace.org

One World UK www.uk.oneworld.net

Oxfam UK Poverty Programme www.oxfamgb.org/ukpp/policy

Peace Direct www.peacedirect.org

Resurgence magazine www.resurgence.org.uk

Royal Society of Arts www.rsa.org.uk

Salt of the Earth www.salt-of-the-earth.org.uk

Social Change And Development (SCAD) www.scadindia.org

The Natural Step www.naturalstep.org

The Web of Hope www.thewebofhope.com.

UK One World Linking Association (UKOWLA) www.ukowla.org.uk

World Changing www.worldchanging.com

World Future Council www.worldfuturecouncil.org

WWF One Planet Living www.wwf.org.uk/oneplanetliving

For further information about The Converging World, please contact us:

Website: www.theconvergingworld.org

Email: info@theconvergingworld.org

OTHER SCHUMACHER BRIEFINGS

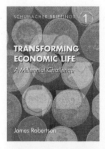   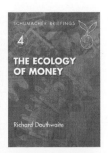

  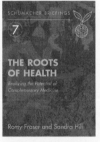 

 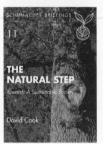 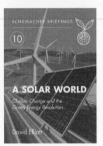 

1. *Transforming Economic Life*
   by James Robertson £5.00

2. *Creating Sustainable Cities*
   by Herbert Girardet £7.00

3. *The Ecology of Health*
   by Robin Stott £5.00

4. *The Ecology of Money*
   by Richard Douthwaite £7.00

5. *Contraction & Convergence*
   by Aubrey Meyer £5.00

6. *Sustainable Education*
   by Stephen Sterling £5.00

7. *The Roots of Health* by Romy Fraser
   and Sandra Hill £5.00

8. *BioRegional Solutions* by Pooran
   Desai and Sue Riddlestone £6.00

9. *Gaian Democracies* by Roy Madron
   and John Jopling £8.00

10. *A Solar World*
    by David Elliott £6.00

11. *The Natural Step*
    by David Cook £6.00

12. *Ecovillages*
    by Jonathan Dawson £6.00